T0094141

Undergraduate Texts in Mathematics

Editors

S. Axler
F.W. Gehring
K.A. Ribet

Springer Science+Business Media, LLC

Undergraduate Texts in Mathematics

Anglin: Mathematics: A Concise History and Philosophy.
Readings in Mathematics.

Anglin/Lambek: The Heritage of Thales.
Readings in Mathematics.

Apostol: Introduction to Analytic Number Theory. Second edition.

Armstrong: Basic Topology.

Armstrong: Groups and Symmetry.

Axler: Linear Algebra Done Right. Second edition.

Beardon: Limits: A New Approach to Real Analysis.

Bak/Newman: Complex Analysis. Second edition.

Banchoff/Wermer: Linear Algebra Through Geometry. Second edition.

Berberian: A First Course in Real Analysis.

Bix: Conics and Cubics: A Concretem Introduction to Algebraic Curves.

Brémaud: An Introduction to Probabilistic Modeling.

Bressoud: Factorization and Primality Testing.

Bressoud: Second Year Calculus.
Readings in Mathematics.

Brickman: Mathematical Introduction to Linear Programming and Game Theory.

Browder: Mathematical Analysis: An Introduction.

Buskes/van Rooij: Topological Spaces: From Distance to Neighborhood.

Cederberg: A Course in Modern Geometries.

Childs: A Concrete Introduction to Higher Algebra. Second edition.

Chung: Elementary Probability Theory with Stochastic Processes. Third edition.

Cox/Little/O'Shea: Ideals, Varieties, and Algorithms. Second edition.

Croom: Basic Concepts of Algebraic Topology.

Curtis: Linear Algebra: An Introductory Approach. Fourth edition.

Devlin: The Joy of Sets: Fundamentals of Contemporary Set Theory. Second edition.

Dixmier: General Topology.

Driver: Why Math?

Ebbinghaus/Flum/Thomas: Mathematical Logic. Second edition.

Edgar: Measure, Topology, and Fractal Geometry.

Elaydi: An Introduction to Difference Equations. Second edition.

Exner: An Accompaniment to Higher Mathematics.

Fine/Rosenberger: The Fundamental Theory of Algebra.

Fischer: Intermediate Real Analysis.

Flanigan/Kazdan: Calculus Two: Linear and Nonlinear Functions. Second edition.

Fleming: Functions of Several Variables. Second edition.

Foulds: Combinatorial Optimization for Undergraduates.

Foulds: Optimization Techniques: An Introduction.

Franklin: Methods of Mathematical Economics.

Frazier: An Introduction to Wavelets Through Linear Algebra.

Gordon: Discrete Probability.

Hairer/Wanner: Analysis by Its History.
Readings in Mathematics.

Halmos: Finite-Dimensional Vector Spaces. Second edition.

Halmos: Naive Set Theory.

Hämmerlin/Hoffmann: Numerical Mathematics.
Readings in Mathematics.

Hartshorne: Geometry: Euclid and Beyond.

Hijab: Introduction to Calculus and Classical Analysis.

Hilton/Holton/Pedersen: Mathematical Reflections: In a Room with Many Mirrors.

Iooss/Joseph: Elementary Stability and Bifurcation Theory. Second edition.

(continued after index)

Joel L. Schiff

The Laplace Transform
Theory and Applications

With 68 Illustrations

 Springer

Joel L. Schiff
Department of Mathematics
The University of Auckland
Private Bag 92019
Auckland
New Zealand
j.schiff@auckland.ac.nz

Mathematics Subject Classification (1991): 44A10

Library of Congress Cataloging-in-Publication Data
Schiff, Joel L.
    The Laplace transform: theory and applications / Joel L. Schiff.
        p.   cm.—(Undergraduate texts in mathematics)
    Includes bibliographical references and index.
    ISBN 978-1-4757-7262-3   ISBN 978-0-387-22757-3 (eBook)
    DOI 10.1007/978-0-387-22757-3
    1. Laplace transformation.   I. Title.   II. Series.
QA432.S33   1999
515'.723—dc21   98-14037

Printed on acid-free paper.

© 1999 Springer Science+Business Media New York
Originally published by Springer-Verlag New York, Inc. in 1999
Softcover reprint of the hardcover 1st edition 1999

Production managed by Robert Bruni; manufacturing supervised by Jeffrey Taub.
Typeset by The Bartlett Press, Inc., Marietta, GA.

9 8 7 6 5 4 3 2 1

ISBN 978-1-4757-7262-3   SPIN 10707361

To my parents

It is customary to begin courses in mathematical engineering by explaining that the lecturer would never trust his life to an aeroplane whose behaviour depended on properties of the Lebesgue integral. It might, perhaps, be just as foolhardy to fly in an aeroplane designed by an engineer who believed that cookbook application of the Laplace transform revealed all that was to be known about its stability.

T.W. Körner
Fourier Analysis
Cambridge University Press
1988

Preface

The Laplace transform is a wonderful tool for solving ordinary and partial differential equations and has enjoyed much success in this realm. With its success, however, a certain casualness has been bred concerning its application, without much regard for hypotheses and when they are valid. Even proofs of theorems often lack rigor, and dubious mathematical practices are not uncommon in the literature for students.

In the present text, I have tried to bring to the subject a certain amount of mathematical correctness and make it accessible to undergraduates. To this end, this text addresses a number of issues that are rarely considered. For instance, when we apply the Laplace transform method to a linear ordinary differential equation with constant coefficients,

$$a_n y^{(n)} + a_{n-1} y^{(n-1)} + \cdots + a_0 y = f(t),$$

why is it justified to take the Laplace transform of both sides of the equation (Theorem A.6)? Or, in many proofs it is required to take the limit inside an integral. This is always frought with danger, especially with an improper integral, and not always justified. I have given complete details (sometimes in the Appendix) whenever this procedure is required.

Furthermore, it is sometimes desirable to take the Laplace transform of an infinite series term by term. Again it is shown that this cannot always be done, and specific sufficient conditions are established to justify this operation.

Another delicate problem in the literature has been the application of the Laplace transform to the so-called Dirac delta function. Except for texts on the theory of distributions, traditional treatments are usually heuristic in nature. In the present text we give a new and mathematically rigorous account of the Dirac delta function based upon the Riemann–Stieltjes integral. It is elementary in scope and entirely suited to this level of exposition.

One of the highlights of the Laplace transform theory is the complex inversion formula, examined in Chapter 4. It is the most sophisticated tool in the Laplace transform arsenal. In order to facilitate understanding of the inversion formula and its many subsequent applications, a self-contained summary of the theory of complex variables is given in Chapter 3.

On the whole, while setting out the theory as explicitly and carefully as possible, the wide range of practical applications for which the Laplace transform is so ideally suited also receive their due coverage. Thus I hope that the text will appeal to students of mathematics and engineering alike.

Historical Summary. Integral transforms date back to the work of Léonard Euler (1763 and 1769), who considered them essentially in the form of the inverse Laplace transform in solving second-order, linear ordinary differential equations. Even Laplace, in his great work, *Théorie analytique des probabilités* (1812), credits Euler with introducing integral transforms. It is Spitzer (1878) who attached the name of *Laplace* to the expression

$$y = \int_a^b e^{sx} \phi(s) \, ds$$

employed by Euler. In this form it is substituted into the differential equation where y is the unknown function of the variable x.

In the late 19th century, the Laplace transform was extended to its complex form by Poincaré and Pincherle, rediscovered by Petzval, and extended to two variables by Picard, with further investigations conducted by Abel and many others.

The first application of the modern Laplace transform occurs in the work of Bateman (1910), who transforms equations arising from Rutherford's work on radioactive decay

$$\frac{dP}{dt} = -\lambda_i P,$$

by setting

$$p(x) = \int_0^\infty e^{-xt} P(t)\, dt$$

and obtaining the transformed equation. Bernstein (1920) used the expression

$$f(s) = \int_0^\infty e^{-su} \phi(u)\, du,$$

calling it the *Laplace transformation,* in his work on theta functions. The modern approach was given particular impetus by Doetsch in the 1920s and 30s; he applied the Laplace transform to differential, integral, and integro-differential equations. This body of work culminated in his foundational 1937 text, *Theorie und Anwendungen der Laplace Transformation.*

No account of the Laplace transformation would be complete without mention of the work of Oliver Heaviside, who produced (mainly in the context of electrical engineering) a vast body of what is termed the "operational calculus." This material is scattered throughout his three volumes, *Electromagnetic Theory* (1894, 1899, 1912), and bears many similarities to the Laplace transform method. Although Heaviside's calculus was not entirely rigorous, it did find favor with electrical engineers as a useful technique for solving their problems. Considerable research went into trying to make the Heaviside calculus rigorous and connecting it with the Laplace transform. One such effort was that of Bromwich, who, among others, discovered the inverse transform

$$X(t) = \frac{1}{2\pi i} \int_{\gamma - i\infty}^{\gamma + i\infty} e^{ts} x(s)\, ds$$

for γ lying to the right of all the singularities of the function x.

Acknowledgments. Much of the Historical Summary has been taken from the many works of Michael Deakin of Monash University. I also wish to thank Alexander Krägeloh for his careful reading of the manuscript and for his many helpful suggestions. I am also indebted to Aimo Hinkkanen, Sergei Federov, Wayne Walker, Nick Dudley Ward, and Allison Heard for their valuable input, to Lev Plimak for the diagrams, to Sione Ma'u for the answers to the exercises, and to Betty Fong for turning my scribbling into a text.

Joel L. Schiff
Auckland
New Zealand

Contents

Preface   **ix**

1 Basic Principles   **1**
  1.1   The Laplace Transform 1
  1.2   Convergence 6
  1.3   Continuity Requirements 8
  1.4   Exponential Order 12
  1.5   The Class $\mathfrak{L}$ 13
  1.6   Basic Properties of the Laplace Transform 16
  1.7   Inverse of the Laplace Transform 23
  1.8   Translation Theorems 27
  1.9   Differentiation and Integration of the
        Laplace Transform 31
  1.10 Partial Fractions 35

2 Applications and Properties   **41**
  2.1   Gamma Function 41
  2.2   Periodic Functions 47
  2.3   Derivatives 53
  2.4   Ordinary Differential Equations 59
  2.5   Dirac Operator 74

xiii

2.6 Asymptotic Values 88
2.7 Convolution 91
2.8 Steady-State Solutions 103
2.9 Difference Equations 108

3 Complex Variable Theory   **115**
3.1 Complex Numbers 115
3.2 Functions 120
3.3 Integration 128
3.4 Power Series 136
3.5 Integrals of the Type $\int_{-\infty}^{\infty} f(x)\,dx$ 147

4 Complex Inversion Formula   **151**

5 Partial Differential Equations   **175**

Appendix   **193**

References   **207**

Tables   **209**
Laplace Transform Operations 209
Table of Laplace Transforms 210

Answers to Exercises   **219**

Index   **231**

1
CHAPTER

Basic Principles

Ordinary and partial differential equations describe the way certain quantities vary with time, such as the current in an electrical circuit, the oscillations of a vibrating membrane, or the flow of heat through an insulated conductor. These equations are generally coupled with initial conditions that describe the state of the system at time $t = 0$.

A very powerful technique for solving these problems is that of the Laplace transform, which literally transforms the original differential equation into an elementary algebraic expression. This latter can then simply be transformed once again, into the solution of the original problem. This technique is known as the "Laplace transform method." It will be treated extensively in Chapter 2. In the present chapter we lay down the foundations of the theory and the basic properties of the Laplace transform.

1.1   The Laplace Transform

Suppose that f is a real- or complex-valued function of the (time) variable $t > 0$ and s is a real or complex parameter. We define the

1

Laplace transform of f as

$$F(s) = \mathcal{L}(f(t)) = \int_0^\infty e^{-st}f(t)\,dt$$

$$= \lim_{\tau \to \infty} \int_0^\tau e^{-st}f(t)\,dt \qquad (1.1)$$

whenever the limit exists (as a finite number). When it does, the integral (1.1) is said to *converge*. If the limit does not exist, the integral is said to *diverge* and there is no Laplace transform defined for f. The notation $\mathcal{L}(f)$ will also be used to denote the Laplace transform of f, and the integral is the ordinary Riemann (improper) integral (see Appendix).

The parameter s belongs to some domain on the real line or in the complex plane. We will choose s appropriately so as to ensure the convergence of the Laplace integral (1.1). In a mathematical and technical sense, the domain of s is quite important. However, in a practical sense, when differential equations are solved, the domain of s is routinely ignored. When s is complex, we will always use the notation $s = x + iy$.

The symbol $\mathcal{L}$ is the *Laplace transformation*, which acts on functions $f = f(t)$ and generates a new function, $F(s) = \mathcal{L}(f(t))$.

Example 1.1. If $f(t) \equiv 1$ for $t \geq 0$, then

$$\mathcal{L}(f(t)) = \int_0^\infty e^{-st}1\,dt$$

$$= \lim_{\tau \to \infty} \left(\frac{e^{-st}}{-s} \Big|_0^\tau \right)$$

$$= \lim_{\tau \to \infty} \left(\frac{e^{-s\tau}}{-s} + \frac{1}{s} \right) \qquad (1.2)$$

$$= \frac{1}{s}$$

provided of course that $s > 0$ (if s is real). Thus we have

$$\mathcal{L}(1) = \frac{1}{s} \qquad (s > 0). \qquad (1.3)$$

If $s \leq 0$, then the integral would diverge and there would be no resulting Laplace transform. If we had taken s to be a complex variable, the same calculation, with $\mathcal{R}e(s) > 0$, would have given $\mathcal{L}(1) = 1/s$.

In fact, let us just verify that in the above calculation the integral can be treated in the same way even if s is a complex variable. We require the well-known Euler formula (see Chapter 3)

$$e^{i\theta} = \cos\theta + i\,\sin\theta, \qquad \theta \text{ real}, \tag{1.4}$$

and the fact that $|e^{i\theta}| = 1$. The claim is that (ignoring the minus sign as well as the limits of integration to simplify the calculation)

$$\int e^{st}\,dt = \frac{e^{st}}{s}, \tag{1.5}$$

for $s = x + iy$ any complex number $\neq 0$. To see this observe that

$$\int e^{st}\,dt = \int e^{(x+iy)t}\,dt$$

$$= \int e^{xt}\cos yt\,dt + i\int e^{xt}\sin yt\,dt$$

by Euler's formula. Performing a double integration by parts on both these integrals gives

$$\int e^{st}\,dt = \frac{e^{xt}}{x^2+y^2}\Big[(x\cos yt + y\sin yt) + i(x\sin yt - y\cos yt)\Big].$$

Now the right-hand side of (1.5) can be expressed as

$$\frac{e^{st}}{s} = \frac{e^{(x+iy)t}}{x+iy}$$

$$= \frac{e^{xt}(\cos yt + i\sin yt)(x - iy)}{x^2 + y^2}$$

$$= \frac{e^{xt}}{x^2+y^2}\Big[(x\cos yt + y\sin yt) + i(x\sin yt - y\cos yt)\Big],$$

which equals the left-hand side, and (1.5) follows.

Furthermore, we obtain the result of (1.3) for s complex if we take $\mathcal{R}e(s) = x > 0$, since then

$$\lim_{\tau\to\infty} |e^{-s\tau}| = \lim_{\tau\to\infty} e^{-x\tau} = 0,$$

killing off the limit term in (1.3).

Let us use the preceding to calculate $\mathcal{L}(\cos \omega t)$ and $\mathcal{L}(\sin \omega t)$ (ω real).

Example 1.2. We begin with

$$
\begin{aligned}
\mathcal{L}(e^{i\omega t}) &= \int_0^\infty e^{-st} e^{i\omega t}\, dt \\
&= \lim_{\tau \to \infty} \frac{e^{(i\omega - s)t}}{i\omega - s}\bigg|_0^\tau \\
&= \frac{1}{s - i\omega},
\end{aligned}
$$

since $\lim_{\tau \to \infty} |e^{i\omega t} e^{-s\tau}| = \lim_{\tau \to \infty} e^{-x\tau} = 0$, provided $x = \mathcal{R}e(s) > 0$. Similarly, $\mathcal{L}(e^{-i\omega t}) = 1/(s + i\omega)$. Therefore, using the linearity property of $\mathcal{L}$, which follows from the fact that integrals are linear operators (discussed in Section 1.6),

$$
\frac{\mathcal{L}(e^{i\omega t}) + \mathcal{L}(e^{-i\omega t})}{2} = \mathcal{L}\left(\frac{e^{i\omega t} + e^{-i\omega t}}{2}\right) = \mathcal{L}(\cos \omega t),
$$

and consequently,

$$
\mathcal{L}(\cos \omega t) = \frac{1}{2}\left(\frac{1}{s - i\omega} + \frac{1}{s + i\omega}\right) = \frac{s}{s^2 + \omega^2}. \tag{1.6}
$$

Similarly,

$$
\mathcal{L}(\sin \omega t) = \frac{1}{2i}\left(\frac{1}{s - i\omega} - \frac{1}{s + i\omega}\right) = \frac{\omega}{s^2 + \omega^2} \qquad (\mathcal{R}e(s) > 0). \tag{1.7}
$$

The Laplace transform of functions defined in a piecewise fashion is readily handled as follows.

Example 1.3. Let (Figure 1.1)

$$
f(t) = \begin{cases} t & 0 \le t \le 1 \\ 1 & t > 1. \end{cases}
$$

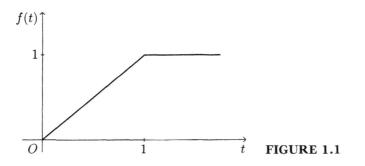

FIGURE 1.1

From the definition,

$$\mathcal{L}(f(t)) = \int_0^\infty e^{-st} f(t)\, dt$$

$$= \int_0^1 te^{-st}\, dt + \lim_{\tau \to \infty} \int_1^\tau e^{-st}\, dt$$

$$= \left.\frac{te^{-st}}{-s}\right|_0^1 + \frac{1}{s}\int_0^1 e^{-st}\, dt + \lim_{\tau \to \infty} \left.\frac{e^{-st}}{-s}\right|_1^\tau$$

$$= \frac{1 - e^{-s}}{s^2} \qquad (\mathcal{R}e(s) > 0).$$

Exercises 1.1

1. From the definition of the Laplace transform, compute $\mathcal{L}(f(t))$ for

(a) $f(t) = 4t$

(b) $f(t) = e^{2t}$

(c) $f(t) = 2\cos 3t$

(d) $f(t) = 1 - \cos \omega t$

(e) $f(t) = te^{2t}$

(f) $f(t) = e^t \sin t$

(g) $f(t) = \begin{cases} 1 & t \geq a \\ 0 & t < a \end{cases}$

(h) $f(t) = \begin{cases} \sin \omega t & 0 < t < \dfrac{\pi}{\omega} \\ 0 & \dfrac{\pi}{\omega} \leq t \end{cases}$

(i) $f(t) = \begin{cases} 2 & t \le 1 \\ e^t & t > 1. \end{cases}$

2. Compute the Laplace transform of the function $f(t)$ whose graph is given in the figures below.

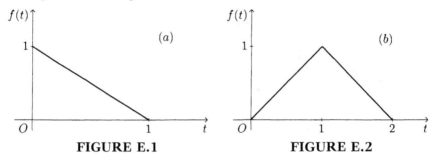

FIGURE E.1   FIGURE E.2

1.2   Convergence

Although the Laplace operator can be applied to a great many functions, there are some for which the integral (1.1) does not converge.

Example 1.4.   For the function $f(t) = e^{(t^2)}$,

$$\lim_{\tau \to \infty} \int_0^\tau e^{-st} e^{t^2} dt = \lim_{\tau \to \infty} \int_0^\tau e^{t^2 - st} dt = \infty$$

for any choice of the variable s, since the integrand grows without bound as $\tau \to \infty$.

In order to go beyond the superficial aspects of the Laplace transform, we need to distinguish two special modes of convergence of the Laplace integral.

The integral (1.1) is said to be *absolutely convergent* if

$$\lim_{\tau \to \infty} \int_0^\tau |e^{-st} f(t)| \, dt$$

exists. If $\mathcal{L}(f(t))$ does converge absolutely, then

$$\left| \int_\tau^{\tau'} e^{-st} f(t) \, dt \right| \le \int_\tau^{\tau'} |e^{-st} f(t)| dt \to 0$$

as $\tau \to \infty$, for all $\tau' > \tau$. This then implies that $\mathcal{L}(f(t))$ also converges in the ordinary sense of (1.1).*

There is another form of convergence that is of the utmost importance from a mathematical perspective. The integral (1.1) is said to *converge uniformly* for s in some domain Ω in the complex plane if for any $\varepsilon > 0$, there exists some number τ_0 such that if $\tau \geq \tau_0$, then

$$\left| \int_\tau^\infty e^{-st} f(t) \, dt \right| < \varepsilon$$

for all s in Ω. The point here is that τ_0 can be chosen sufficiently large in order to make the "tail" of the integral arbitrarily small, *independent of* s.

Exercises 1.2

1. Suppose that f is a continuous function on $[0, \infty)$ and $|f(t)| \leq M < \infty$ for $0 \leq t < \infty$.

   (a) Show that the Laplace transform $F(s) = \mathcal{L}(f(t))$ converges absolutely (and hence converges) for any s satisfying $\mathcal{R}e(s) > 0$.
   (b) Show that $\mathcal{L}(f(t))$ converges uniformly if $\mathcal{R}e(s) \geq x_0 > 0$.
   (c) Show that $F(s) = \mathcal{L}(f(t)) \to 0$ as $\mathcal{R}e(s) \to \infty$.

2. Let $f(t) = e^t$ on $[0, \infty)$.

   (a) Show that $F(s) = \mathcal{L}(e^t)$ converges for $\mathcal{R}e(s) > 1$.
   (b) Show that $\mathcal{L}(e^t)$ converges uniformly if $\mathcal{R}e(s) \geq x_0 > 1$.

---

*Convergence of an integral

$$\int_0^\infty \varphi(t) \, dt$$

is equivalent to the *Cauchy criterion*:

$$\int_\tau^{\tau'} \varphi(t) dt \to 0 \qquad \text{as} \qquad \tau \to \infty, \ \tau' > \tau.$$

(c) Show that $F(s) = \mathcal{L}(e^t) \to 0$ as $\mathcal{R}e(s) \to \infty$.

3. Show that the Laplace transform of the function $f(t) = 1/t$, $t > 0$ does not exist for any value of s.

1.3 Continuity Requirements

Since we can compute the Laplace transform for some functions and not others, such as $e^{(t^2)}$, we would like to know that there is a large class of functions that do have a Laplace tranform. There is such a class once we make a few restrictions on the functions we wish to consider.

Definition 1.5. A function f has a **jump discontinuity at a point** t_0 if both the limits

$$\lim_{t \to t_0^-} f(t) = f(t_0^-) \qquad \text{and} \qquad \lim_{t \to t_0^+} f(t) = f(t_0^+)$$

exist (as finite numbers) and $f(t_0^-) \neq f(t_0^+)$. Here, $t \to t_0^-$ and $t \to t_0^+$ mean that $t \to t_0$ from the left and right, respectively (Figure 1.2).

Example 1.6. The function (Figure 1.3)

$$f(t) = \frac{1}{t - 3}$$

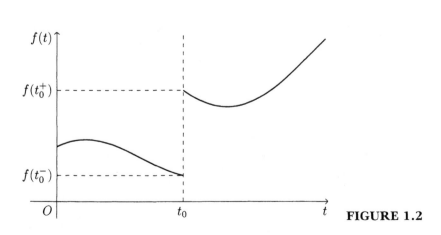

FIGURE 1.2

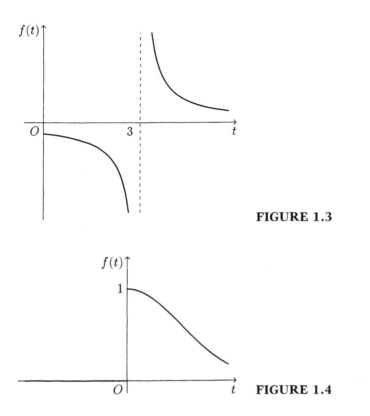

FIGURE 1.3

FIGURE 1.4

has a discontinuity at $t = 3$, but it is not a jump discontinuity since neither $\lim_{t \to 3^-} f(t)$ nor $\lim_{t \to 3^+} f(t)$ exists.

Example 1.7. The function (Figure 1.4)

$$f(t) = \begin{cases} e^{-\frac{t^2}{2}} & t > 0 \\ 0 & t < 0 \end{cases}$$

has a jump discontinuity at $t = 0$ and is continuous elsewhere.

Example 1.8. The function (Figure 1.5)

$$f(t) = \begin{cases} 0 & t < 0 \\ \cos \frac{1}{t} & t > 0 \end{cases}$$

is discontinuous at $t = 0$, but $\lim_{t \to 0^+} f(t)$ fails to exist, so f does not have a jump discontinuity at $t = 0$.

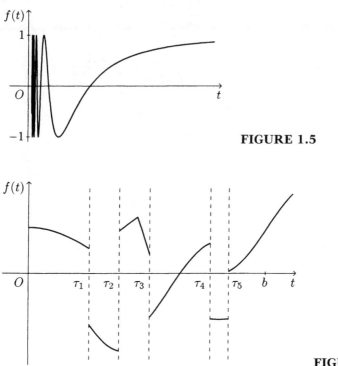

FIGURE 1.5

FIGURE 1.6

The class of functions for which we consider the Laplace transform defined will have the following property.

Definition 1.9. A function f is **piecewise continuous** on the interval $[0, \infty)$ if (i) $\lim_{t \to 0^+} f(t) = f(0^+)$ exists and (ii) f is continuous on every finite interval $(0, b)$ except possibly at a finite number of points $\tau_1, \tau_2, \ldots, \tau_n$ in $(0, b)$ at which f has a jump discontinuity (Figure 1.6).

The function in Example 1.6 is not piecewise continuous on $[0, \infty)$. Nor is the function in Example 1.8. However, the function in Example 1.7 is piecewise continuous on $[0, \infty)$.

An important consequence of piecewise continuity is that on each subinterval the function f is also *bounded*. That is to say,

$$|f(t)| \le M_i, \quad \tau_i < t < \tau_{i+1}, \quad i = 1, 2, \ldots, n-1,$$

for finite constants M_i.

In order to integrate piecewise continuous functions from 0 to b, one simply integrates f over each of the subintervals and takes the sum of these integrals, that is,

$$\int_0^b f(t)\,dt = \int_0^{\tau_1} f(t)\,dt + \int_{\tau_1}^{\tau_2} f(t)\,dt + \cdots + \int_{\tau_n}^b f(t)\,dt.$$

This can be done since the function f is both continuous and bounded on each subinterval and thus on each has a well-defined (Riemann) integral.

Exercises 1.3

Discuss the continuity of each of the following functions and locate any jump discontinuities.

1. $f(t) = \dfrac{1}{1+t}$

2. $g(t) = t \sin \dfrac{1}{t} \quad (t \neq 0)$

3. $h(t) = \begin{cases} t & t \leq 1 \\ \dfrac{1}{1+t^2} & t > 1 \end{cases}$

4. $i(t) = \begin{cases} \dfrac{\sinh t}{t} & t \neq 0 \\ 1 & t = 0 \end{cases}$

5. $j(t) = \dfrac{1}{t} \sinh \dfrac{1}{t} \quad (t \neq 0)$

6. $k(t) = \begin{cases} \dfrac{1 - e^{-t}}{t} & t \neq 0 \\ 0 & t = 0 \end{cases}$

7. $l(t) = \begin{cases} 1 & 2na \leq t < (2n+1)a \\ -1 & (2n+1)a \leq t < (2n+2)a \end{cases} \qquad a > 0,\, n = 0,\, 1,\, 2,\, \dots$

8. $m(t) = \left[\dfrac{t}{a} \right] + 1$, for $t \geq 0$, $a > 0$, where $[x] = $ greatest integer $\leq x$.

1.4   Exponential Order

The second consideration of our class of functions possessing a well-defined Laplace transform has to do with the growth rate of the functions. In the definition

$$\mathcal{L}(f(t)) = \int_0^\infty e^{-st} f(t)\, dt,$$

when we take $s > 0$ (or $\mathcal{R}e(s) > 0$), the integral will converge as long as f does not grow too rapidly. We have already seen by Example 1.4 that $f(t) = e^{t^2}$ does grow too rapidly for our purposes. A suitable rate of growth can be made explicit.

Definition 1.10.   A function f has **exponential order** α if there exist constants $M > 0$ and α such that for some $t_0 \geq 0$,

$$|f(t)| \leq M\, e^{\alpha t}, \qquad t \geq t_0.$$

Clearly the exponential function e^{at} has exponential order $\alpha = a$, whereas t^n has exponential order α for any $\alpha > 0$ and any $n \in \mathbb{N}$ (Exercises 1.4, Question 2), and bounded functions like $\sin t$, $\cos t$, $\tan^{-1} t$ have exponential order 0, whereas e^{-t} has order -1. However, e^{t^2} does not have exponential order. Note that if $\beta > \alpha$, then exponential order α implies exponential order β, since $e^{\alpha t} \leq e^{\beta t}$, $t \geq 0$. We customarily state the order as the smallest value of α that works, and if the value itself is not significant it may be suppressed altogether.

Exercises 1.4

1. If f_1 and f_2 are piecewise continuous functions of orders α and β, respectively, on $[0, \infty)$, what can be said about the continuity and order of the functions

   (i) $c_1 f_1 + c_2 f_2$, c_1, c_2 constants,
   (ii) $f \cdot g$?

2. Show that $f(t) = t^n$ has exponential order α for any $\alpha > 0$, $n \in \mathbb{N}$.
3. Prove that the function $g(t) = e^{t^2}$ does not have exponential order.

1.5 The Class $\mathfrak{L}$

We now show that a large class of functions possesses a Laplace transform.

Theorem 1.11. *If f is piecewise continuous on $[0, \infty)$ and of exponential order α, then the Laplace transform $\mathcal{L}(f)$ exists for $\mathcal{R}e(s) > \alpha$ and converges absolutely.*

PROOF. First,

$$|f(t)| \leq M_1 \, e^{\alpha t}, \qquad t \geq t_0,$$

for some real α. Also, f is piecewise continuous on $[0, t_0]$ and hence bounded there (the bound being just the largest bound over all the subintervals), say

$$|f(t)| \leq M_2, \qquad 0 < t < t_0.$$

Since $e^{\alpha t}$ has a positive minimum on $[0, t_0]$, a constant M can be chosen sufficiently large so that

$$|f(t)| \leq M \, e^{\alpha t}, \qquad t > 0.$$

Therefore,

$$\int_0^\tau |e^{-st} f(t)| dt \leq M \int_0^\tau e^{-(x-\alpha)t} dt$$

$$= \frac{M \, e^{-(x-\alpha)t}}{-(x-\alpha)} \Big|_0^\tau$$

$$= \frac{M}{x-\alpha} - \frac{M \, e^{-(x-\alpha)\tau}}{x-\alpha}.$$

Letting $\tau \to \infty$ and noting that $\mathcal{R}e(s) = x > \alpha$ yield

$$\int_0^\infty |e^{-st} f(t)| dt \leq \frac{M}{x-\alpha}. \tag{1.8}$$

Thus the Laplace integral converges absolutely in this instance (and hence converges) for $\mathcal{R}e(s) > \alpha$. □

Example 1.12. Let $f(t) = e^{at}$, a real. This function is continuous on $[0, \infty)$ and of exponential order a. Then

$$\mathcal{L}(e^{at}) = \int_0^\infty e^{-st} e^{at} dt$$

$$= \int_0^\infty e^{-(s-a)t} dt$$

$$= \frac{e^{-(s-a)t}}{-(s-a)} \Big|_0^\infty = \frac{1}{s-a} \qquad (\mathcal{R}e(s) > a).$$

The same calculation holds for a complex and $\mathcal{R}e(s) > \mathcal{R}e(a)$.

Example 1.13. Applying integration by parts to the function $f(t) = t$ $(t \geq 0)$, which is continuous and of exponential order, gives

$$\mathcal{L}(t) = \int_0^\infty t e^{-st} dt$$

$$= \frac{-t e^{-st}}{s} \Big|_0^\infty + \frac{1}{s} \int_0^\infty e^{-st} dt$$

$$= \frac{1}{s} \mathcal{L}(1) \qquad (\text{provided } \mathcal{R}e(s) > 0)$$

$$= \frac{1}{s^2}.$$

Performing integration by parts twice as above, we find that

$$\mathcal{L}(t^2) = \int_0^\infty e^{-st} t^2 dt$$

$$= \frac{2}{s^3} \qquad (\mathcal{R}e(s) > 0).$$

By induction, one can show that in general,

$$\mathcal{L}(t^n) = \frac{n!}{s^{n+1}} \qquad (\mathcal{R}e(s) > 0) \tag{1.9}$$

for $n = 1, 2, 3, \ldots$. Indeed, this formula holds even for $n = 0$, since $0! = 1$, and will be shown to hold even for non-integer values of n in Section 2.1.

Let us define the class L as the set of those real- or complex-valued functions defined on the open interval $(0, \infty)$ for which the

Laplace transform (defined in terms of the Riemann integral) exists for some value of s. It is known that whenever $F(s) = \mathcal{L}(f(t))$ exists for some value s_0, then $F(s)$ exists for all s with $\mathcal{R}e(s) > \mathcal{R}e(s_0)$, that is, the Laplace transform exists for all s in some right half-plane (cf. Doetsch [2], Theorem 3.4). By Theorem 1.11, piecewise continuous functions on $[0, \infty)$ having exponential order belong to L. However, there certainly are functions in L that do not satisfy one or both of these conditions.

Example 1.14. Consider

$$f(t) = 2t\, e^{t^2} \cos(e^{t^2}).$$

Then $f(t)$ is continuous on $[0, \infty)$ but not of exponential order. However, the Laplace transform of $f(t)$,

$$\mathcal{L}(f(t)) = \int_0^\infty e^{-st} 2t\, e^{t^2} \cos(e^{t^2}) dt,$$

exists, since integration by parts yields

$$\mathcal{L}(f(t)) = e^{-st} \sin(e^{t^2})\Big|_0^\infty + s \int_0^\infty e^{-st} \sin(e^{t^2})\, dt$$

$$= -\sin(1) + s\, \mathcal{L}(\sin(e^{t^2})) \qquad (\mathcal{R}e(s) > 0).$$

and the latter Laplace transform exists by Theorem 1.11. Thus we have a continuous function that is not of exponential order yet nevertheless possesses a Laplace transform. See also Remark 2.8.

Another example is the function

$$f(t) = \frac{1}{\sqrt{t}}. \tag{1.10}$$

We will compute its actual Laplace transform in Section 2.1 in the context of the gamma function. While (1.10) has exponential order $\alpha = 0$ $\left(|f(t)| \le 1,\ t \ge 1\right)$, it is not piecewise continuous on $[0, \infty)$ since $f(t) \to \infty$ as $t \to 0^+$, that is, $t = 0$ is not a jump discontinuity.

Exercises 1.5

1. Consider the function $g(t) = t\, e^{t^2} \sin(e^{t^2})$.

(a) Is g continuous on $[0, \infty)$? Does g have exponential order?
(b) Show that the Laplace transform $F(s)$ exists for $\mathcal{R}e(s) > 0$.
(c) Show that g is the derivative of some function having exponential order.

2. Without actually determining it, show that the following functions possess a Laplace transform.

(a) $\dfrac{\sin t}{t}$

(b) $\dfrac{1 - \cos t}{t}$

(c) $t^2 \sinh t$

3. Without determining it, show that the function f, whose graph is given in Figure E.3, possesses a Laplace transform. (See Question 3(a), Exercises 1.7.)

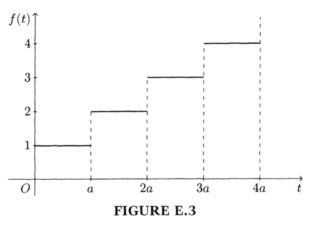

FIGURE E.3

1.6 Basic Properties of the Laplace Transform

Linearity. One of the most basic and useful properties of the Laplace operator $\mathcal{L}$ is that of *linearity*, namely, if $f_1 \in L$ for $\mathcal{R}e(s) > \alpha$, $f_2 \in L$ for $\mathcal{R}e(s) > \beta$, then $f_1 + f_2 \in L$ for $\mathcal{R}e(s) > \max\{\alpha, \beta\}$, and

$$\mathcal{L}(c_1 f_1 + c_2 f_2) = c_1 \mathcal{L}(f_1) + c_2 \mathcal{L}(f_2) \tag{1.11}$$

for arbitrary constants c_1, c_2.

This follows from the fact that integration is a linear process, to wit,

$$\int_0^\infty e^{-st}\left(c_1 f_1(t) + c_2 f_2(t)\right) dt$$

$$= c_1 \int_0^\infty e^{-st} f_1(t)\, dt + c_2 \int_0^\infty e^{-st} f_2(t)\, dt \qquad (f_1, f_2 \in L).$$

Example 1.15. The hyperbolic cosine function

$$\cosh \omega t = \frac{e^{\omega t} + e^{-\omega t}}{2}$$

describes the curve of a hanging cable between two supports. By linearity

$$\mathcal{L}(\cosh \omega t) = \frac{1}{2}[\mathcal{L}(e^{\omega t}) + \mathcal{L}(e^{-\omega t})]$$

$$= \frac{1}{2}\left(\frac{1}{s - \omega} + \frac{1}{s + \omega}\right)$$

$$= \frac{s}{s^2 - \omega^2}.$$

Similarly,

$$\mathcal{L}(\sinh \omega t) = \frac{\omega}{s^2 - \omega^2}.$$

Example 1.16. If $f(t) = a_0 + a_1 t + \cdots + a_n t^n$ is a polynomial of degree n, then

$$\mathcal{L}\bigl(f(t)\bigr) = \sum_{k=0}^n a_k \mathcal{L}(t^k)$$

$$= \sum_{k=0}^n \frac{a_k k!}{s^{k+1}}$$

by (1.9) and (1.11).

Infinite Series. For an infinite series, $\sum_{n=0}^\infty a_n t^n$, in general it is not possible to obtain the Laplace transform of the series by taking the transform term by term.

Example 1.17.

$$f(t) = e^{-t^2} = \sum_{n=0}^{\infty} \frac{(-1)^n t^{2n}}{n!}, \qquad -\infty < t < \infty.$$

Taking the Laplace transform term by term gives

$$\sum_{n=0}^{\infty} \frac{(-1)^n}{n!} \mathcal{L}(t^{2n}) = \sum_{n=0}^{\infty} \frac{(-1)^n}{n!} \frac{(2n)!}{s^{2n+1}}$$

$$= \frac{1}{s} \sum_{n=0}^{\infty} \frac{(-1)^n (2n) \cdots (n+2)(n+1)}{s^{2n}}.$$

Applying the ratio test,

$$\lim_{n \to \infty} \left| \frac{u_{n+1}}{u_n} \right| = \lim_{n \to \infty} \frac{2(2n+1)}{|s|^2} = \infty,$$

and so the series diverges for all values of s.

However, $\mathcal{L}(e^{-t^2})$ does exist since e^{-t^2} is continuous and bounded on $[0, \infty)$.

So when can we guarantee obtaining the Laplace transform of an infinite series by term-by-term computation?

Theorem 1.18. *If*

$$f(t) = \sum_{n=0}^{\infty} a_n t^n$$

converges for $t \geq 0$, with

$$|a_n| \leq \frac{K\alpha^n}{n!},$$

for all n sufficiently large and $\alpha > 0$, $K > 0$, then

$$\mathcal{L}(f(t)) = \sum_{n=0}^{\infty} a_n \mathcal{L}(t^n) = \sum_{n=0}^{\infty} \frac{a_n n!}{s^{n+1}} \qquad (\mathcal{R}e(s) > \alpha).$$

PROOF.   Since $f(t)$ is represented by a convergent power series, it is continuous on $[0, \infty)$. We desire to show that the difference

$$\left| \mathcal{L}(f(t)) - \sum_{n=0}^{N} a_n \mathcal{L}(t^n) \right| = \left| \mathcal{L}\left(f(t) - \sum_{n=0}^{N} a_n t^n \right) \right|$$

$$\le \mathcal{L}_x\left(\left|f(t) - \sum_{n=0}^{N} a_n t^n\right|\right)$$

converges to zero as $N \to \infty$, where $\mathcal{L}_x(h(t)) = \int_0^\infty e^{-xt} h(t)\, dt$, $x = \mathcal{R}e(s)$.

To this end,

$$\left|f(t) - \sum_{n=0}^{N} a_n t^n\right| = \left|\sum_{n=N+1}^{\infty} a_n t^n\right|$$

$$\le K \sum_{n=N+1}^{\infty} \frac{(\alpha t)^n}{n!}$$

$$= K\left(e^{\alpha t} - \sum_{n=0}^{N} \frac{(\alpha t)^n}{n!}\right)$$

since $e^x = \sum_{n=0}^{\infty} x^n/n!$. As $h \le g$ implies $\mathcal{L}_x(h) \le \mathcal{L}_x(g)$ when the transforms exist,

$$\mathcal{L}_x\left(\left|f(t) - \sum_{n=0}^{N} a_n t^n\right|\right) \le K\mathcal{L}_x\left(e^{\alpha t} - \sum_{n=0}^{N} \frac{(\alpha t)^n}{n!}\right)$$

$$= K\left(\frac{1}{x - \alpha} - \sum_{n=0}^{N} \frac{\alpha^n}{x^{n+1}}\right)$$

$$= K\left(\frac{1}{x - \alpha} - \frac{1}{x}\sum_{n=0}^{N} \left(\frac{\alpha}{x}\right)^n\right)$$

$$\to 0 \qquad \left(\mathcal{R}e(s) = x > \alpha\right)$$

as $N \to \infty$. We have used the fact that the geometric series has the sum

$$\sum_{n=0}^{\infty} z^n = \frac{1}{1 - z}, \qquad |z| < 1.$$

Therefore,

$$\mathcal{L}(f(t)) = \lim_{N \to \infty} \sum_{n=0}^{N} a_n \mathcal{L}(t^n).$$

$$= \sum_{n=0}^{\infty} \frac{a_n n!}{s^{n+1}} \qquad (\mathcal{R}e(s) > \alpha). \qquad \Box$$

Note that the coefficients of the series in Example 1.17 do not satisfy the hypothesis of the theorem.

Example 1.19.

$$f(t) = \frac{\sin t}{t} = \sum_{n=0}^{\infty} \frac{(-1)^n t^{2n}}{(2n+1)!}.$$

Then,

$$|a_{2n}| = \frac{1}{(2n+1)!} < \frac{1}{(2n)!}, \qquad n = 0, 1, 2, \ldots,$$

and so we can apply the theorem:

$$\mathcal{L}\left(\frac{\sin t}{t}\right) = \sum_{n=0}^{\infty} \frac{(-1)^n \mathcal{L}(t^{2n})}{(2n+1)!}$$

$$= \sum_{n=0}^{\infty} \frac{(-1)^n}{(2n+1)s^{2n+1}}$$

$$= \tan^{-1}\left(\frac{1}{s}\right), \qquad |s| > 1.$$

Here we are using the fact that

$$\tan^{-1} x = \int_0^x \frac{dt}{1+t^2} = \int_0^x \sum_{n=0}^{\infty} (-1)^n t^{2n}$$

$$= \sum_{n=0}^{\infty} \frac{(-1)^n x^{2n+1}}{2n+1}, \qquad |x| < 1,$$

with $x = 1/s$, as we can integrate the series term by term. See also Example 1.38.

Uniform Convergence. We have already seen by Theorem 1.11 that for functions f that are piecewise continuous on $[0, \infty)$ and of exponential order, the Laplace integral converges absolutely, that is, $\int_0^{\infty} |e^{-st} f(t)| \, dt$ converges. Moreover, for such functions the Laplace integral *converges uniformly*.

To see this, suppose that

$$|f(t)| \leq M \, e^{\alpha t}, \qquad t \geq t_0.$$

Then

$$\left| \int_{t_0}^{\infty} e^{-st} f(t) \, dt \right| \leq \int_{t_0}^{\infty} e^{-xt} |f(t)| dt$$

$$\leq M \int_{t_0}^{\infty} e^{-(x-\alpha)t} dt$$

$$= \frac{M \, e^{-(x-\alpha)t}}{-(x-\alpha)} \Big|_{t_0}^{\infty}$$

$$= \frac{M \, e^{-(x-\alpha)t_0}}{x-\alpha},$$

provided $x = \mathcal{R}e(s) > \alpha$. Taking $x \geq x_0 > \alpha$ gives an upper bound for the last expression:

$$\frac{M \, e^{-(x-\alpha)t_0}}{x-\alpha} \leq \frac{M}{x_0 - \alpha} e^{-(x_0-\alpha)t_0}. \tag{1.12}$$

By choosing t_0 sufficiently large, we can make the term on the right-hand side of (1.12) arbitrarily small; that is, given any $\varepsilon > 0$, there exists a value $T > 0$ such that

$$\left| \int_{t_0}^{\infty} e^{-st} f(t) \, dt \right| < \varepsilon, \qquad \text{whenever} \quad t_0 \geq T \tag{1.13}$$

for all values of s with $\mathcal{R}e(s) \geq x_0 > \alpha$. This is precisely the condition required for the uniform convergence of the Laplace integral in the region $\mathcal{R}e(s) \geq x_0 > \alpha$ (see Section 1.2). The importance of the uniform convergence of the Laplace transform cannot be overemphasized, as it is instrumental in the proofs of many results.

$F(s) \to \Gamma$ **as** $s \to \infty$. A general property of the Laplace transform that becomes apparent from an inspection of the table at the back of this book (pp. 210–218) is the following.

Theorem 1.20. *If f is piecewise continuous on $[0, \infty)$ and has exponential order α, then*

$$F(s) = \mathcal{L}(f(t)) \to 0$$

as $\mathcal{R}e(s) \to \infty$.

In fact, by (1.8)

$$\left| \int_0^\infty e^{-st} f(t) \, dt \right| \leq \frac{M}{x - \alpha}, \qquad (\mathcal{R}e(s) = x > \alpha),$$

and letting $x \to \infty$ gives the result.

Remark 1.21. As it turns out, $F(s) \to 0$ as $\mathcal{R}e(s) \to \infty$ whenever the Laplace transform exists, that is, for all $f \in L$ (cf. Doetsch [2], Theorem 23.2). As a consequence, any function $F(s)$ without this behavior, say $(s - 1)/(s + 1)$, e^s/s, or s^2, cannot be the Laplace transform of any function f.

Exercises 1.6

1. Find $\mathcal{L}(2t + 3e^{2t} + 4\sin 3t)$.
2. Show that $\mathcal{L}(\sinh \omega t) = \dfrac{\omega}{s^2 - \omega^2}$.
3. Compute

   (a) $\mathcal{L}(\cosh^2 \omega t)$ (b) $\mathcal{L}(\sinh^2 \omega t)$.

4. Find $\mathcal{L}(3\cosh 2t - 2\sinh 2t)$.
5. Compute $\mathcal{L}(\cos \omega t)$ and $\mathcal{L}(\sin \omega t)$ from the Taylor series representations

   $$\cos \omega t = \sum_{n=0}^\infty \frac{(-1)^n (\omega t)^{2n}}{(2n)!}, \qquad \sin \omega t = \sum_{n=0}^\infty \frac{(-1)^n (\omega t)^{2n+1}}{(2n + 1)!},$$

   respectively.
6. Determine $\mathcal{L}(\sin^2 \omega t)$ and $\mathcal{L}(\cos^2 \omega t)$ using the formulas

   $$\sin^2 \omega t = \frac{1}{2} - \frac{1}{2}\cos 2\omega t, \qquad \cos^2 \omega t = 1 - \sin^2 \omega t,$$

   respectively.
7. Determine $\mathcal{L}\left(\dfrac{1 - e^{-t}}{t} \right)$.

Hint:

$$\log(1 + x) = \sum_{n=0}^{\infty} \frac{(-1)^n x^{n+1}}{n+1}, \qquad |x| < 1.$$

8. Determine $\mathcal{L}\left(\dfrac{1 - \cos \omega t}{t}\right)$.

9. Can $F(s) = s/\log s$ be the Laplace transform of some function f?

1.7 Inverse of the Laplace Transform

In order to apply the Laplace transform to physical problems, it is necessary to invoke the inverse transform. If $\mathcal{L}(f(t)) = F(s)$, then the *inverse Laplace transform* is denoted by

$$\mathcal{L}^{-1}(F(s)) = f(t), \qquad t \geq 0,$$

which maps the Laplace transform of a function back to the original function. For example,

$$\mathcal{L}^{-1}\left(\frac{\omega}{s^2 + \omega^2}\right) = \sin \omega t, \qquad t \geq 0.$$

The question naturally arises: Could there be some other function $f(t) \neq \sin \omega t$ with $\mathcal{L}^{-1}(\omega/(s^2 + \omega^2)) = f(t)$? More generally, we need to know when the inverse transform is *unique*.

Example 1.22. Let

$$g(t) = \begin{cases} \sin \omega t & t > 0 \\ 1 & t = 0. \end{cases}$$

Then

$$\mathcal{L}(g(t)) = \frac{\omega}{s^2 + \omega^2},$$

since altering a function at a single point (or even at a finite number of points) does not alter the value of the Laplace (Riemann) integral.

This example illustrates that $\mathcal{L}^{-1}(F(s))$ can be more than one function, in fact infinitely many, at least when considering functions with discontinuities. Fortunately, this is the only case (cf. Doetsch [2], p. 24).

Theorem 1.23. *Distinct continuous functions on $[0, \infty)$ have distinct Laplace transforms.*

This result is known as *Lerch's theorem*. It means that if we restrict our attention to functions that are continuous on $[0, \infty)$, then the inverse transform

$$\mathcal{L}^{-1}(F(s)) = f(t)$$

is uniquely defined and we can speak about *the* inverse, $\mathcal{L}^{-1}(F(s))$. This is exactly what we shall do in the sequel, and hence we write

$$\mathcal{L}^{-1}\left(\frac{\omega}{s^2 + \omega^2}\right) = \sin \omega t, \qquad t \geq 0.$$

Since many of the functions we will be dealing with will be solutions to differential equations and hence continuous, the above assumptions are completely justified.

Note also that $\mathcal{L}^{-1}$ is *linear*, that is,

$$\mathcal{L}^{-1}(a\,F(s) + b\,G(s)) = a\,f(t) + b\,g(t)$$

if $\mathcal{L}(f(t)) = F(s)$, $\mathcal{L}(g(t)) = G(s)$. This follows from the linearity of $\mathcal{L}$ and holds in the domain common to F and G.

Example 1.24.

$$\mathcal{L}^{-1}\left(\frac{1}{2(s-1)} + \frac{1}{2(s+1)}\right) = \frac{1}{2}e^t + \frac{1}{2}e^{-t}$$
$$= \cosh t, \qquad t \geq 0.$$

One of the practical features of the Laplace transform is that it can be applied to *discontinuous* functions f. In these instances, it must be borne in mind that when the inverse transform is invoked, there are other functions with the same $\mathcal{L}^{-1}(F(s))$.

Example 1.25. An important function occurring in electrical systems is the (*delayed*) *unit step function* (Figure 1.7)

$$u_a(t) = \begin{cases} 1 & t \geq a \\ 0 & t < a, \end{cases}$$

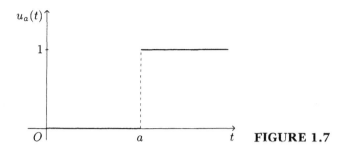

FIGURE 1.7

for $a \geq 0$. This function delays its output until $t = a$ and then assumes a constant value of one unit. In the literature, the unit step function is also commonly defined as

$$u_a(t) = \begin{cases} 1 & t > a \\ 0 & t < a, \end{cases}$$

for $a \geq 0$, and is known as the *Heaviside (step) function*. Both definitions of $u_a(t)$ have the same Laplace transform and so from that point of view are indistinguishable. When $a = 0$, we will write $u_a(t) = u(t)$. Another common notation for the unit step function $u_a(t)$ is $u(t - a)$.

Computing the Laplace transform,

$$\mathcal{L}\big(u_a(t)\big) = \int_0^\infty e^{-st} u_a(t) \, dt$$

$$= \int_a^\infty e^{-st} \, dt$$

$$= \frac{e^{-st}}{-s} \Big|_a^\infty$$

$$= \frac{e^{-as}}{s} \qquad \big(\mathcal{R}e(s) > 0\big).$$

It is appropriate to write $\big($with either interpretation of $u_a(t)\big)$

$$\mathcal{L}^{-1}\left(\frac{e^{-as}}{s}\right) = u_a(t),$$

although we could equally have written $\mathcal{L}^{-1}\left(e^{-as}/s\right) = v_a(t)$ for

$$v_a(t) = \begin{cases} 1 & t > a \\ 0 & t \leq a, \end{cases}$$

which is another variant of the unit step function.

Another interesting function along these lines is the following.

Example 1.26. For $0 \leq a < b$, let

$$u_{ab}(t) = \frac{1}{b-a}\left(u_a(t) - u_b(t)\right) = \begin{cases} 0 & t < a \\ \frac{1}{b-a} & a \leq t < b \\ 0 & t \geq b, \end{cases}$$

as shown in Figure 1.8.

Then

$$\mathcal{L}\left(u_{ab}(t)\right) = \frac{e^{-as} - e^{-bs}}{s(b-a)}.$$

Exercises 1.7

1. Prove that $\mathcal{L}^{-1}$ is a linear operator.

2. A function $N(t)$ is called a *null function* if

$$\int_0^t N(\tau)\,d\tau = 0,$$

for all $t > 0$.

(a) Give an example of a null function that is not identically zero.

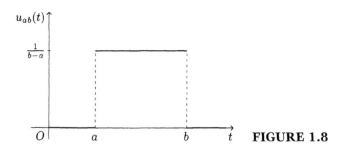

FIGURE 1.8

(b) Use integration by parts to show that

$$\mathcal{L}(N(t)) = 0,$$

for any null function $N(t)$.

(c) Conclude that

$$\mathcal{L}(f(t) + N(t)) = \mathcal{L}(f(t)),$$

for any $f \in L$ and null function $N(t)$. (The converse is also true, namely, if $\mathcal{L}(f_1) \equiv \mathcal{L}(f_2)$ in a right half-plane, then f_1 and f_2 differ by at most a null function. See Doetsch [2], pp. 20–24).

(d) How can part (c) be reconciled with Theorem 1.23?

3. Consider the function f whose graph is given in Question 3 of Exercises 1.5 (Figure E.3).

(a) Compute the Laplace transform of f directly from the explicit values $f(t)$ and deduce that

$$\mathcal{L}(f(t)) = \frac{1}{s(1 - e^{-as})} \qquad (\mathcal{R}e(s) > 0, a > 0).$$

(b) Write $f(t)$ as an infinite series of unit step functions.

(c) By taking the Laplace transform term by term of the infinite series in (b), show that the same result as in (a) is attained.

1.8   Translation Theorems

We present two very useful results for determining Laplace transforms and their inverses. The first pertains to a translation in the s-domain and the second to a translation in the t-domain.

Theorem 1.27 (First Translation Theorem).   *If $F(s) = \mathcal{L}(f(t))$ for $\mathcal{R}e(s) > 0$, then*

$$F(s - a) = \mathcal{L}(e^{at} f(t)) \qquad (a \text{ real}, \ \mathcal{R}e(s) > a).$$

PROOF.   For $\mathcal{R}e(s) > a$,

$$F(s - a) = \int_0^\infty e^{-(s-a)t} f(t) \, dt$$

$$= \int_0^\infty e^{-st} e^{at} f(t)\, dt$$

$$= \mathcal{L}\big(e^{at} f(t)\big). \qquad\qquad \square$$

Example 1.28. Since

$$\mathcal{L}(t) = \frac{1}{s^2} \qquad (\mathcal{R}e(s) > 0),$$

then

$$\mathcal{L}(t\, e^{at}) = \frac{1}{(s-a)^2} \qquad (\mathcal{R}e(s) > a),$$

and in general,

$$\mathcal{L}(t^n e^{at}) = \frac{n!}{(s-a)^{n+1}}, \qquad n = 0, 1, 2, \ldots \qquad (\mathcal{R}e(s) > a).$$

This gives a useful inverse:

$$\mathcal{L}^{-1}\left(\frac{1}{(s-a)^{n+1}}\right) = \frac{1}{n!}\, t^n e^{at}, \qquad t \ge 0.$$

Example 1.29. Since

$$\mathcal{L}(\sin \omega t) = \frac{\omega}{s^2 + \omega^2},$$

then

$$\mathcal{L}(e^{2t} \sin 3t) = \frac{3}{(s-2)^2 + 9}.$$

In general,

$$\mathcal{L}(e^{at} \cos \omega t) = \frac{s-a}{(s-a)^2 + \omega^2} \qquad (\mathcal{R}e(s) > a)$$

$$\mathcal{L}(e^{at} \sin \omega t) = \frac{\omega}{(s-a)^2 + \omega^2} \qquad (\mathcal{R}e(s) > a)$$

$$\mathcal{L}(e^{at} \cosh \omega t) = \frac{s-a}{(s-a)^2 - \omega^2} \qquad (\mathcal{R}e(s) > a)$$

$$\mathcal{L}(e^{at} \sinh \omega t) = \frac{\omega}{(s-a)^2 - \omega^2} \qquad (\mathcal{R}e(s) > a).$$

Example 1.30.

$$\mathcal{L}^{-1}\left(\frac{s}{s^2 + 4s + 1}\right) = \mathcal{L}^{-1}\left(\frac{s}{(s+2)^2 - 3}\right)$$

$$= \mathcal{L}^{-1}\left(\frac{s+2}{(s+2)^2 - 3}\right) - \mathcal{L}^{-1}\left(\frac{2}{(s+2)^2 - 3}\right)$$

$$= e^{-2t}\cosh\sqrt{3}t - \frac{2}{\sqrt{3}}e^{-2t}\sinh\sqrt{3}t.$$

In the first step we have used the procedure of *completing the square*.

Theorem 1.31 (Second Translation Theorem). *If $F(s) = \mathcal{L}(f(t))$, then*

$$\mathcal{L}\left(u_a(t)f(t-a)\right) = e^{-as}F(s) \qquad (a \geq 0).$$

This follows from the basic fact that

$$\int_0^\infty e^{-st}[u_a(t)f(t-a)]\,dt = \int_a^\infty e^{-st}f(t-a)\,dt,$$

and setting $\tau = t - a$, the right-hand integral becomes

$$\int_0^\infty e^{-s(\tau+a)}f(\tau)\,d\tau = e^{-as}\int_0^\infty e^{-st}f(\tau)\,d\tau$$

$$= e^{-as}F(s).$$

Example 1.32. Let us determine $\mathcal{L}(g(t))$ for (Figure 1.9)

$$g(t) = \begin{cases} 0 & 0 \leq t < 1 \\ (t-1)^2 & t \geq 1. \end{cases}$$

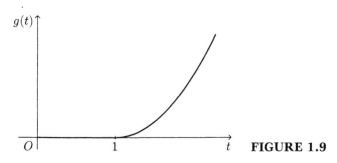

FIGURE 1.9

Note that $g(t)$ is just the function $f(t) = t^2$ delayed by $(a =) 1$ unit of time. Whence

$$\begin{aligned}
\mathcal{L}(g(t)) &= \mathcal{L}\left(u_1(t)(t-1)^2\right) \\
&= e^{-s}\mathcal{L}(t^2) \\
&= \frac{2e^{-s}}{s^3} \qquad (\mathcal{R}e(s) > 0).
\end{aligned}$$

The second translation theorem can also be considered in inverse form:

$$\mathcal{L}^{-1}\left(e^{-as}F(s)\right) = u_a(t)f(t-a), \qquad (1.14)$$

for $F(s) = \mathcal{L}(f(t))$, $a \geq 0$.

Example 1.33.   Find

$$\mathcal{L}^{-1}\left(\frac{e^{-2s}}{s^2+1}\right).$$

We have

$$\frac{e^{-2s}}{s^2+1} = e^{-2s}\mathcal{L}(\sin t),$$

so by (1.14)

$$\mathcal{L}^{-1}\left(\frac{e^{-2s}}{s^2+1}\right) = u_2(t)\sin(t-2), \qquad (t \geq 0).$$

This is just the function $\sin t$, which gets "turned on" at time $t = 2$.

Exercises 1.8

1. Determine

    (a) $\mathcal{L}(e^{2t}\sin 3t)$ **(b)** $\mathcal{L}(t^2 e^{-\omega t})$

    (c) $\mathcal{L}^{-1}\left(\dfrac{4}{(s-4)^3}\right)$ **(d)** $\mathcal{L}(e^{7t}\sinh\sqrt{2}\,t)$

    (e) $\mathcal{L}^{-1}\left(\dfrac{1}{s^2+2s+5}\right)$ **(f)** $\mathcal{L}^{-1}\left(\dfrac{s}{s^2+6s+1}\right)$

(g) $\mathcal{L}\left(e^{-at}\cos(\omega t + \theta)\right)$ **(h)** $\mathcal{L}^{-1}\left(\dfrac{s}{(s+1)^2}\right).$

2. Determine $\mathcal{L}(f(t))$ for

(a) $f(t) = \begin{cases} 0 & 0 \le t < 2 \\ e^{at} & t \ge 2 \end{cases}$ **(b)** $f(t) = \begin{cases} 0 & 0 \le t < \frac{\pi}{2} \\ \sin t & t \ge \frac{\pi}{2} \end{cases}$

(c) $f(t) = u_\pi(t)\cos(t - \pi).$

3. Find

(a) $\mathcal{L}^{-1}\left(\dfrac{e^{-2s}}{s^3}\right)$

(b) $\mathcal{L}^{-1}\left(\dfrac{E}{s} - \dfrac{s}{s^2 + 1}e^{-as}\right)$ (E constant)

(c) $\mathcal{L}^{-1}\left(\dfrac{e^{-\pi s}}{s^2 - 2}\right).$

1.9 Differentiation and Integration of the Laplace Transform

As will be shown in Chapter 3, when s is a complex variable, the Laplace transform $F(s)$ (for suitable functions) is an analytic function of the parameter s. When s is a real variable, we have a formula for the derivative of $F(s)$, which holds in the complex case as well (Theorem 3.3).

Theorem 1.34. *Let f be piecewise continuous on $[0, \infty)$ of exponential order α and $\mathcal{L}(f(t)) = F(s)$. Then*

$$\frac{d^n}{ds^n} F(s) = \mathcal{L}\left((-1)^n t^n f(t)\right), \qquad n = 1, 2, 3, \ldots \ (s > \alpha). \qquad (1.15)$$

PROOF. By virtue of the hypotheses, for $s \ge x_0 > \alpha$, it is justified (cf. Theorem A.12) to interchange the derivative and integral sign in the following calculation.

$$\frac{d}{ds} F(s) = \frac{d}{ds} \int_0^\infty e^{-st} f(t)\, dt$$

$$= \int_0^\infty \frac{\partial}{\partial s} e^{-st} f(t)\, dt$$

$$= \int_0^\infty -t e^{-st} f(t)\, dt$$

$$= \mathcal{L}\big(- t f(t)\big).$$

Since for any $s > \alpha$, one can find some x_0 satisfying $s \geq x_0 > \alpha$, the preceding result holds for any $s > \alpha$. Repeated differentiation (or rather induction) gives the general case, by virtue of $\mathcal{L}\big(t^k f(t)\big)$ being uniformly convergent for $s \geq x_0 > \alpha$. $\square$

Example 1.35.

$$\mathcal{L}(t \cos \omega t) = -\frac{d}{ds} \mathcal{L}(\cos \omega t)$$

$$= -\frac{d}{ds} \frac{s}{s^2 + \omega^2}$$

$$= \frac{s^2 - \omega^2}{(s^2 + \omega^2)^2}.$$

Similarly,

$$\mathcal{L}(t \sin \omega t) = \frac{2\omega s}{(s^2 + \omega^2)^2}.$$

For $n = 1$ we can express (1.15) as

$$f(t) = -\frac{1}{t} \mathcal{L}^{-1} \left(\frac{d}{ds} F(s) \right) \qquad (t > 0) \tag{1.16}$$

for $f(t) = \mathcal{L}^{-1}\big(F(s)\big)$. This formulation is also useful.

Example 1.36. Find

$$f(t) = \mathcal{L}^{-1} \left(\log \frac{s+a}{s+b} \right).$$

Since

$$\frac{d}{ds} \log \left(\frac{s+a}{s+b} \right) = \frac{1}{s+a} - \frac{1}{s+b},$$

$$f(t) = -\frac{1}{t} \mathcal{L}^{-1} \left(\frac{1}{s+a} - \frac{1}{s+b} \right)$$

$$= \frac{1}{t}(e^{-bt} - e^{-at}).$$

Not only can the Laplace transform be differentiated, but it can be integrated as well. Again the result is another Laplace transform.

Theorem 1.37. *If f is piecewise continuous on $[0, \infty)$ and of exponential order α, with $F(s) = \mathcal{L}(f(t))$ and such that $\lim_{t \to 0+} f(t)/t$ exists, then*

$$\int_s^\infty F(x)\,dx = \mathcal{L}\left(\frac{f(t)}{t} \right) \qquad (s > \alpha).$$

PROOF.   Integrating both sides of the equation

$$F(x) = \int_0^\infty e^{-xt} f(t)\,dt \qquad (x \text{ real}),$$

we obtain

$$\int_s^\infty F(x)\,dx = \lim_{w \to \infty} \int_s^w \left(\int_0^\infty e^{-xt} f(t)\,dt \right) dx.$$

As $\int_0^\infty e^{-xt} f(t)\,dt$ converges uniformly for $\alpha < s \le x \le w$ (1.12), we can reverse the order of integration (cf. Theorem A.11), giving

$$\int_s^\infty F(x)\,dx = \lim_{w \to \infty} \int_0^\infty \left(\int_s^w e^{-xt} f(t)\,dx \right) dt$$

$$= \lim_{w \to \infty} \int_0^\infty \left[\frac{e^{-xt}}{-t} f(t) \right]_s^w dt$$

$$= \int_0^\infty e^{-st} \frac{f(t)}{t}\,dt - \lim_{w \to \infty} \int_0^\infty e^{-wt} \frac{f(t)}{t}\,dt$$

$$= \mathcal{L}\left(\frac{f(t)}{t} \right),$$

as $\lim_{w \to \infty} G(w) = 0$ by Theorem 1.20 for $G(w) = \mathcal{L}\left(f(t)/t\right)$. The existence of $\mathcal{L}\left(f(t)/t\right)$ is ensured by the hypotheses. ☐

Example 1.38.

(i) $\quad \mathcal{L}\left(\dfrac{\sin t}{t}\right) = \displaystyle\int_s^\infty \dfrac{dx}{x^2 + 1} = \dfrac{\pi}{2} - \tan^{-1} s$

$\qquad\qquad = \tan^{-1}\left(\dfrac{1}{s}\right) \qquad (s > 0).$

(ii) $\quad \mathcal{L}\left(\dfrac{\sinh \omega t}{t}\right) = \displaystyle\int_s^\infty \dfrac{\omega\, dx}{x^2 - \omega^2}$

$\qquad\qquad = \dfrac{1}{2}\displaystyle\int_s^\infty \left(\dfrac{1}{x - \omega} - \dfrac{1}{x + \omega}\right) dx$

$\qquad\qquad = \dfrac{1}{2} \ln \dfrac{s + \omega}{s - \omega} \qquad (s > |\omega|).$

Exercises 1.9

1. Determine

   (a) $\mathcal{L}(t \cosh \omega t)$ 　　　　　　　　　(b) $\mathcal{L}(t \sinh \omega t)$

   (c) $\mathcal{L}(t^2 \cos \omega t)$ 　　　　　　　　(d) $\mathcal{L}(t^2 \sin \omega t)$.

2. Using Theorem 1.37, show that

   (a) $\mathcal{L}\left(\dfrac{1 - e^{-t}}{t}\right) = \log\left(1 + \dfrac{1}{s}\right) \quad (s > 0)$

   (b) $\mathcal{L}\left(\dfrac{1 - \cos \omega t}{t}\right) = \tfrac{1}{2}\log\left(1 + \dfrac{\omega^2}{s^2}\right) \quad (s > 0).$

   [Compare (a) and (b) with Exercises 1.6, Question 7 and 8, respectively.]

   (c) $\mathcal{L}\left(\dfrac{1 - \cosh \omega t}{t}\right) = \dfrac{1}{2}\log\left(1 - \dfrac{\omega^2}{s^2}\right) \quad (s > |\omega|).$

3. Using (1.16), find

   (a) $\mathcal{L}^{-1}\left(\log\left(\dfrac{s^2 + a^2}{s^2 + b^2}\right)\right)$ 　　　　(b) $\mathcal{L}^{-1}\left(\tan^{-1}\dfrac{1}{s}\right) \quad (s > 0).$

4. If

$$\mathcal{L}^{-1}\left(\frac{e^{-a\sqrt{s}}}{\sqrt{s}}\right) = \frac{e^{-a^2/4t}}{\sqrt{\pi t}},$$

find $\mathcal{L}^{-1}(e^{-a\sqrt{s}})$.

1.10 Partial Fractions

In many applications of the Laplace transform it becomes necessary to find the inverse of a particular transform, $F(s)$. Typically it is a function that is not immediately recognizable as the Laplace transform of some elementary function, such as

$$F(s) = \frac{1}{(s-2)(s-3)},$$

for s confined to some region Ω (e.g., $\mathcal{R}e(s) > \alpha$). Just as in calculus (for s real), where the goal is to integrate such a function, the procedure required here is to decompose the function into *partial fractions*.

In the preceding example, we can decompose $F(s)$ into the sum of two fractional expressions:

$$\frac{1}{(s-2)(s-3)} = \frac{A}{s-2} + \frac{B}{s-3},$$

that is,

$$1 = A(s-3) + B(s-2). \tag{1.17}$$

Since (1.17) equates two polynomials [1 and $A(s-3) + B(s-2)$] that are equal for all s in Ω, except possibly for $s = 2$ and $s = 3$, the two polynomials are identically equal for all values of s. This follows from the fact that two polynomials of degree n that are equal at more than n points are identically equal (Corollary A.8).

Thus, if $s = 2$, $A = -1$, and if $s = 3$, $B = 1$, so that

$$F(s) = \frac{1}{(s-2)(s-3)} = \frac{-1}{s-2} + \frac{1}{s-3}.$$

Finally,

$$f(t) = \mathcal{L}^{-1}(F(s)) = \mathcal{L}^{-1}\left(-\frac{1}{s-2}\right) + \mathcal{L}^{-1}\left(\frac{1}{s-3}\right)$$

$$= -e^{2t} + e^{3t}.$$

Partial Fraction Decompositions. We will be concerned with the quotient of two polynomials, namely a rational function

$$F(s) = \frac{P(s)}{Q(s)},$$

where the degree of $Q(s)$ is greater than the degree of $P(s)$, and $P(s)$ and $Q(s)$ have no common factors. Then $F(s)$ can be expressed as a finite sum of partial fractions.

(i) For each linear factor of the form $as + b$ of $Q(s)$, there corresponds a partial fraction of the form

$$\frac{A}{as+b}, \qquad A \text{ constant.}$$

(ii) For each repeated linear factor of the form $(as + b)^n$, there corresponds a partial fraction of the form

$$\frac{A_1}{as+b} + \frac{A_2}{(as+b)^2} + \cdots + \frac{A_n}{(as+b)^n}, \qquad A_1, A_2, \ldots, A_n \text{ constants.}$$

(iii) For every quadratic factor of the form $as^2 + bs + c$, there corresponds a partial fraction of the form

$$\frac{As+B}{as^2+bs+c}, \qquad A, B \text{ constants.}$$

(iv) For every repeated quadratic factor of the form $(as^2+bs+c)^n$, there corresponds a partial fraction of the form

$$\frac{A_1 s + B_1}{as^2+bs+c} + \frac{A_2 s + B_2}{(as^2+bs+c)^2} + \cdots + \frac{A_n s + B_n}{(as^2+bs+c)^n},$$

$$A_1, \ldots, A_n, B_1, \ldots, B_n \text{ constants.}$$

The object is to determine the constants once the polynomial $P(s)/Q(s)$ has been represented by a partial fraction decomposition. This can be achieved by several different methods.

Example 1.39.

$$\frac{1}{(s-2)(s-3)} = \frac{A}{s-2} + \frac{B}{s-3}$$

or

$$1 = A(s-3) + B(s-2),$$

as we have already seen. Since this is a polynomial identity valid for all s, we may equate the coefficients of like powers of s on each side of the equals sign (see Corollary A.8). Thus, for s, $0 = A + B$; and for s^0, $1 = -3A - 2B$. Solving these two equations simultaneously, $A = -1$, $B = 1$ as before.

Example 1.40. Find

$$\mathcal{L}^{-1}\left(\frac{s+1}{s^2(s-1)}\right).$$

Write

$$\frac{s+1}{s^2(s-1)} = \frac{A}{s} + \frac{B}{s^2} + \frac{C}{s-1},$$

or

$$s+1 = As(s-1) + B(s-1) + Cs^2,$$

which is an identity for all values of s. Setting $s = 0$ gives $B = -1$; setting $s = 1$ gives $C = 2$. Equating the coefficients of s^2 gives $0 = A + C$, and so $A = -2$. Whence

$$\mathcal{L}^{-1}\left(\frac{s+1}{s^2(s-1)}\right) = -2\mathcal{L}^{-1}\left(\frac{1}{s}\right) - \mathcal{L}^{-1}\left(\frac{1}{s^2}\right) + 2\mathcal{L}^{-1}\left(\frac{1}{s-1}\right)$$

$$= -2 - t + 2e^t.$$

Example 1.41. Find

$$\mathcal{L}^{-1}\left(\frac{2s^2}{(s^2+1)(s-1)^2}\right).$$

We have

$$\frac{2s^2}{(s^2+1)(s-1)^2} = \frac{As+B}{s^2+1} + \frac{C}{s-1} + \frac{D}{(s-1)^2},$$

or

$$2s^2 = (As + B)(s - 1)^2 + C(s^2 + 1)(s - 1) + D(s^2 + 1).$$

Setting $s = 1$ gives $D = 1$. Also, setting $s = 0$ gives $0 = B - C + D$, or

$$-1 = B - C.$$

Equating coefficients of s^3 and s, respectively,

$$0 = A + C,$$
$$0 = A - 2B + C.$$

These last two equations imply $B = 0$. Then from the first equation, $C = 1$; finally, the second equation shows $A = -1$. Therefore,

$$\mathcal{L}^{-1}\left(\frac{2s^2}{(s^2 + 1)(s - 1)^2}\right) = -\mathcal{L}^{-1}\left(\frac{s}{s^2 + 1}\right) + \mathcal{L}^{-1}\left(\frac{1}{s - 1}\right)$$

$$+ \mathcal{L}^{-1}\left(\frac{1}{(s - 1)^2}\right)$$

$$= -\cos t + e^t + te^t.$$

Simple Poles. Suppose that we have $F(t) = \mathcal{L}(f(t))$ for

$$F(s) = \frac{P(s)}{Q(s)} = \frac{P(s)}{(s - \alpha_1)(s - \alpha_2) \cdots (s - \alpha_n)}, \qquad \alpha_i \neq \alpha_j,$$

where $P(s)$ is a polyomial of degree less than n. In the terminology of complex variables (cf. Chapter 3), the α_is are known as *simple poles* of $F(s)$. A partial fraction decomposition is

$$F(s) = \frac{A_1}{s - \alpha_1} + \frac{A_2}{s - \alpha_2} + \cdots + \frac{A_n}{s - \alpha_n}. \tag{1.18}$$

Multiplying both sides of (1.18) by $s - \alpha_i$ and letting $s \to \alpha_i$ yield

$$A_i = \lim_{s \to \alpha_i}(s - \alpha_i)F(s). \tag{1.19}$$

(In Chapter 3 we will see that the A_is are the *residues* of $F(s)$ at the poles α_i.) Therefore,

$$f(t) = \mathcal{L}^{-1}(F(s)) = \sum_{i=1}^{n}\mathcal{L}^{-1}\left(\frac{A_i}{s - \alpha_i}\right) = \sum_{i=1}^{n}A_i e^{\alpha_i t}.$$

Putting in the expression (1.19) for A_i gives a quick method for finding the inverse:

$$f(t) = \mathcal{L}^{-1}\big(F(s)\big) = \sum_{i=1}^{n} \lim_{s \to \alpha_i} (s - \alpha_i)\, F(s)\, e^{\alpha_i t}. \qquad (1.20)$$

Example 1.42. Find

$$\mathcal{L}^{-1}\left(\frac{s}{(s-1)(s+2)(s-3)}\right).$$

$$f(t) = \lim_{s \to 1}(s-1)F(s)e^{t} + \lim_{s \to -2}(s+2)F(s)e^{-2t} + \lim_{s \to 3}(s-3)F(s)e^{3t}$$

$$= -\frac{1}{6}e^{t} - \frac{2}{15}e^{-2t} + \frac{3}{10}e^{3t}.$$

Exercises 1.10

1. Find $\mathcal{L}^{-1}$ of the following transforms $F(s)$ by the partial fraction method.

(a) $\dfrac{1}{(s-a)(s-b)}$

(b) $\dfrac{s}{2s^2 + s - 1}$

(c) $\dfrac{s^2 + 1}{s(s-1)^3}$

(d) $\dfrac{s}{(s^2 + a^2)(s^2 + b^2)}$ $\quad (a \neq b)$

(e) $\dfrac{s}{(s^2 + a^2)(s^2 - b^2)}$

(f) $\dfrac{s+2}{s^5 - 3s^4 + 2s^3}$

(g) $\dfrac{2s^2 + 3}{(s+1)^2(s^2+1)^2}$

(h) $\dfrac{s^2 + s + 3}{s(s^3 - 6s^2 + 5s + 12)}$

(See Example 2.42).

2. Determine

$$\mathcal{L}^{-1}\left(\frac{s^2}{(s^2 - a^2)(s^2 - b^2)(s^2 - c^2)}\right)$$

(a) by the partial fraction method
(b) by using (1.20).

Applications and Properties

The various types of problems that can be treated with the Laplace transform include ordinary and partial differential equations as well as integral and integro-differential equations. In this chapter we delineate the principles of the *Laplace transform method* for the purposes of solving all but PDEs (which we discuss in Chapter 5).

In order to expand our repetoire of Laplace transforms, we discuss the gamma function, periodic functions, infinite series, convolutions, as well as the Dirac delta function, which is not really a function at all in the conventional sense. This latter is considered in an entirely new but rigorous fashion from the standpoint of the Riemann–Stieltjes integral.

2.1 Gamma Function

Recall from equation (1.9) that

$$\mathcal{L}(t^n) = \frac{n!}{s^{n+1}}, \qquad n = 1, 2, 3, \ldots.$$

In order to extend this result for non-integer values of n, consider

$$\mathcal{L}(t^\nu) = \int_0^\infty e^{-st} t^\nu dt \qquad (\nu > -1).$$

Actually, for $-1 < \nu < 0$, the function $f(t) = t^\nu$ is not piecewise continuous on $[0, \infty)$ since it becomes infinite as $t \to 0^+$. However, as the (improper) integral $\int_0^\tau t^\nu dt$ exists for $\nu > -1$, and $f(t) = t^\nu$ is bounded for all large values of t, the Laplace transform, $\mathcal{L}(t^\nu)$, exists.

By a change of variables, $x = st$ $(s > 0)$,

$$\mathcal{L}(t^\nu) = \int_0^\infty e^{-x} \left(\frac{x}{s}\right)^\nu \frac{1}{s} dx$$

$$= \frac{1}{s^{\nu+1}} \int_0^\infty x^\nu e^{-x} dx. \qquad (2.1)$$

The quantity

$$\Gamma(p) = \int_0^\infty x^{p-1} e^{-x} dx \qquad (p > 0)$$

is known as the (*Euler*) *gamma function*. Although the improper integral exists and is a continuous function of $p > 0$, it is not equal to any elementary function (Figure 2.1).

Then (2.1) becomes

$$\mathcal{L}(t^\nu) = \frac{\Gamma(\nu+1)}{s^{\nu+1}}, \qquad \nu > -1, \ s > 0. \qquad (2.2)$$

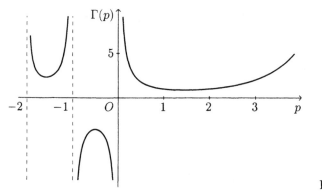

FIGURE 2.1

Comparing (1.9) with (2.2) when $v = n = 0, 1, 2, \ldots$ yields

$$\Gamma(n+1) = n!. \tag{2.3}$$

Thus we see that the gamma function is a generalization of the notion of factorial. In fact, it can be defined for all complex values of v, $v \neq 0, -1, -2, \cdots$, and enjoys the factorial property

$$\Gamma(v+1) = v\Gamma(v), \qquad v \neq 0, -1, -2, \ldots$$

(see Exercises 2.1, Question 1).

Example 2.1. For $v = -1/2$,

$$\mathcal{L}\left(t^{-\frac{1}{2}}\right) = \frac{\Gamma\left(\frac{1}{2}\right)}{s^{\frac{1}{2}}},$$

where

$$\Gamma\left(\tfrac{1}{2}\right) = \int_0^\infty x^{-\frac{1}{2}} e^{-x} \, dx.$$

Making a change of variables, $x = u^2$,

$$\Gamma\left(\tfrac{1}{2}\right) = 2 \int_0^\infty e^{-u^2} \, du.$$

This integral is well known in the theory of probability and has the value $\sqrt{\pi}$. (To see this, write

$$I^2 = \left(\int_0^\infty e^{-x^2} \, dx\right)\left(\int_0^\infty e^{-y^2} \, dy\right) = \int_0^\infty \int_0^\infty e^{-(x^2+y^2)} \, dx \, dy,$$

and evaluate the double integral by polar coordinates, to get $I = \sqrt{\pi}/2$.)
Hence

$$\mathcal{L}\left(t^{-\frac{1}{2}}\right) = \sqrt{\frac{\pi}{s}} \qquad (s > 0) \tag{2.4}$$

and

$$\mathcal{L}^{-1}\left(s^{-\frac{1}{2}}\right) = \frac{1}{\sqrt{\pi t}} \qquad (t > 0). \tag{2.5}$$

Example 2.2. Determine

$$\mathcal{L}(\log t) = \int_0^\infty e^{-st} \log t \, dt.$$

Again setting $x = st$, $s > 0$,

$$\mathcal{L}(\log t) = \int_0^\infty e^{-x} \log\left(\frac{x}{s}\right)\frac{1}{s}\,dx$$

$$= \frac{1}{s}\left(\int_0^\infty e^{-x}\log x\,dx - \log s \int_0^\infty e^{-x}dx\right)$$

$$= -\frac{1}{s}(\log s + \gamma), \tag{2.6}$$

where

$$\gamma = -\int_0^\infty e^{-x}\log x\,dx = 0.577215\ldots$$

is *Euler's constant*. See also Exercises 2.1, Question 4.

Infinite Series. If

$$f(t) = \sum_{n=0}^\infty a_n t^{n+\nu} \qquad (\nu > -1)$$

converges for all $t \geq 0$ and $|a_n| \leq K(\alpha^n/n!)$, K, $\alpha > 0$, for all n sufficiently large, then

$$\mathcal{L}(f(t)) = \sum_{n=0}^\infty \frac{a_n \Gamma(n+\nu+1)}{s^{n+\nu+1}} \qquad (\mathcal{R}e(s) > \alpha).$$

This generalizes Theorem 1.18 (cf. Watson [14], P 1.3.1). In terms of the inverse transform, if

$$F(s) = \sum_{n=0}^\infty \frac{a_n}{s^{n+\nu+1}} \qquad (\nu > -1), \tag{2.7}$$

where the series converges for $|s| > R$, then the inverse can be computed term by term:

$$f(t) = \mathcal{L}^{-1}(F(s)) = \sum_{n=0}^\infty \frac{a_n}{\Gamma(n+\nu+1)}t^{n+\nu}, \qquad t \geq 0. \tag{2.8}$$

To verify (2.8), note that since the series in (2.7) converges for $|s| > R$,

$$\left|\frac{a_n}{s^n}\right| \leq K$$

for some constant K and for all n. Then for $|s| = r > R$,

$$|a_n| \leq K r^n. \tag{2.9}$$

Also,

$$r^n < \frac{2^n}{n} r^n = \frac{\alpha^n}{n}, \tag{2.10}$$

taking $\alpha = 2r$. Since $\Gamma(n+v+1) \geq \Gamma(n)$ for $v > -1$, $n \geq 2$, (2.9) and (2.10) imply

$$\frac{|a_n|}{\Gamma(n+v+1)} \leq \frac{K \alpha^n}{n \Gamma(n)} = \frac{K \alpha^n}{n!}, \tag{2.11}$$

as required.

Furthermore, (2.11) guarantees

$$\left| \frac{a_n}{\Gamma(n+v+1)} \right| t^n \leq \frac{K(\alpha t)^n}{n!} \qquad (t \geq 0),$$

and as $\sum_{n=0}^{\infty} (\alpha t)^n / n! = e^{\alpha t}$ converges, (2.8) converges absolutely. This also shows that f has exponential order.

Taking $v = 0$ in (2.7): If

$$F(s) = \sum_{n=0}^{\infty} \frac{a_n}{s^{n+1}}$$

converges for $|s| > R$, then the inverse is given by

$$f(t) = \mathcal{L}^{-1}(F(s)) = \sum_{n=0}^{\infty} \frac{a_n}{n!} t^n.$$

Example 2.3. Suppose

$$F(s) = \frac{1}{\sqrt{s+a}} = \frac{1}{\sqrt{s}} \left(1 + \frac{a}{s} \right)^{-\frac{1}{2}} \qquad (a \text{ real}).$$

Using the binomial series expansion for $(1+x)^\alpha$,

$$F(s) = \frac{1}{\sqrt{s}} \left[1 - \frac{1}{2} \left(\frac{a}{s} \right) + \frac{\left(\frac{1}{2} \right) \left(\frac{3}{2} \right)}{2!} \left(\frac{a}{s} \right)^2 - \frac{\left(\frac{1}{2} \right) \left(\frac{3}{2} \right) \left(\frac{5}{2} \right)}{3!} \left(\frac{a}{s} \right)^3 \right.$$
$$\left. + \cdots + \frac{(-1)^n \cdot 1 \cdot 3 \cdot 5 \cdots (2n-1)}{2^n n!} \left(\frac{a}{s} \right)^n + \cdots \right]$$

$$= \sum_{n=0}^{\infty} \frac{(-1)^n \cdot 1 \cdot 3 \cdot 5 \cdots (2n-1)a^n}{2^n n! \, s^{n+\frac{1}{2}}}, \qquad |s| > |a|.$$

Inverting in accordance with (2.8),

$$f(t) = \mathcal{L}^{-1}\left(F(s)\right) = \sum_{n=0}^{\infty} \frac{(-1)^n 1 \cdot 3 \cdot 5 \cdots (2n-1)a^n t^{n-\frac{1}{2}}}{2^n n! \, \Gamma\left(n+\frac{1}{2}\right)}$$

$$= \frac{1}{\sqrt{t}} \sum_{n=0}^{\infty} \frac{(-1)^n 1 \cdot 3 \cdot 5 \cdots (2n-1)a^n t^n}{2^n n! \, \Gamma\left(n+\frac{1}{2}\right)}.$$

Here we can use the formula $v\,\Gamma(v) = \Gamma(v+1)$ to find by induction that

$$\Gamma\left(n+\frac{1}{2}\right) = \Gamma\left(\frac{1}{2}\right)\left(\frac{1 \cdot 3 \cdot 5 \cdots (2n-1)}{2^n}\right)$$

$$= \sqrt{\pi}\left(\frac{1 \cdot 3 \cdot 5 \cdots (2n-1)}{2^n}\right).$$

Thus

$$f(t) = \frac{1}{\sqrt{t}} \sum_{n=0}^{\infty} \frac{(-1)^n a^n t^n}{\sqrt{\pi}\, n!}$$

$$= \frac{1}{\sqrt{\pi t}} \, e^{-at}.$$

Note that in this case $f(t)$ can also be determined from the first translation theorem (1.27) and (2.5).

Exercises 2.1

1. Establish the "factorial property" of the gamma function

$$\Gamma(v+1) = v\,\Gamma(v),$$

   for $v > 0$.
2. Compute

   (a) $\Gamma\left(\frac{3}{2}\right)$
   (b) $\Gamma(3)$

(c) $\Gamma\left(-\frac{1}{2}\right)$ **(d)** $\Gamma\left(-\frac{3}{2}\right)$.

3. Compute

(a) $\mathcal{L}\left(\dfrac{e^{3t}}{\sqrt{t}}\right)$ **(b)** $\mathcal{L}^{-1}\left(\dfrac{e^{-2s}}{\sqrt{s}}\right)$

(c) $\mathcal{L}^{-1}\left(\dfrac{1}{(s-a)^{3/2}}\right)$ **(d)** $\mathcal{L}^{-1}\left(\displaystyle\sum_{n=0}^{\infty}\dfrac{(-1)^{n}}{s^{n+1}}\right)$, $|s| > 1$

(e) $\mathcal{L}^{-1}\left(\displaystyle\sum_{n=1}^{\infty}\dfrac{(-1)^{n+1}}{ns^{2n}}\right)$, $|s| > 1$

(f) $\mathcal{L}(\sqrt{t})$.

4. (a) Show that

$$\frac{\partial}{\partial v}t^{v-1} = t^{v-1}\log t.$$

(b) From (a) and 2.2 prove that

$$\mathcal{L}(t^{v-1}\log t) = \frac{\Gamma'(v) - \Gamma(v)\log s}{s^{v}}, \qquad s > 0, \;\; v > 0.$$

(c) Conclude that

$$\mathcal{L}(\log t) = -\frac{1}{s}(\log s + \gamma),$$

where

$$\gamma = -\int_{0}^{\infty} e^{-x}\log x\,dx = 0.577215\ldots,$$

is the Euler constant as in (2.6).

2.2 Periodic Functions

If a function f is *periodic* with period $T > 0$, then $f(t) = f(t + T)$, $-\infty < t < \infty$. The periodic functions $\sin t$ and $\cos t$ both have period

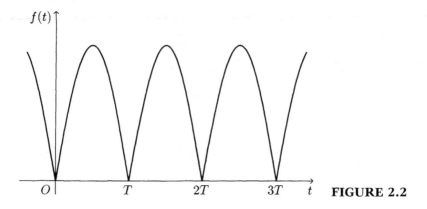

FIGURE 2.2

$T = 2\pi$, whereas $\tan t$ has period $T = \pi$. Since the functions f with which we are dealing are defined only for $t \geq 0$, we adopt the same condition for periodicity as above for these functions as well.

The function f in Figure 2.2, is periodic with period T. We define

$$F_1(s) = \int_0^T e^{-st} f(t)\, dt, \qquad (2.12)$$

which is the Laplace transform of the function denoting the first period and zero elsewhere.

The Laplace transform of the entire function f is just a particular multiple of this first one.

Theorem 2.4. *If* $F(s) = \mathcal{L}(f(t))$ *and* f *is periodic of period* T, *then*

$$F(s) = \frac{1}{1 - e^{-sT}} F_1(s). \qquad (2.13)$$

PROOF.

$$F(s) = \int_0^\infty e^{-st} f(t)\, dt = \int_0^T e^{-st} f(t)\, dt + \int_T^\infty e^{-st} f(t)\, dt.$$

Changing variables with $\tau = t - T$ in the last integral,

$$\int_T^\infty e^{-st} f(t)\, dt = \int_0^\infty e^{-s(\tau+T)} f(\tau + T)\, d\tau$$

$$= e^{-sT} \int_0^\infty e^{-s\tau} f(\tau)\, d\tau$$

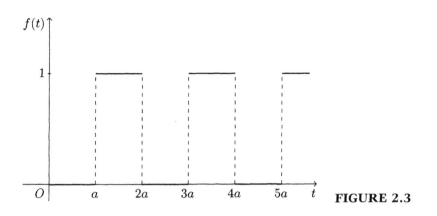

FIGURE 2.3

by the periodicity of f. Therefore,

$$F(s) = \int_0^T e^{-st} f(t)\, dt + e^{-sT} F(s);$$

solving,

$$F(s) = \frac{1}{1 - e^{-sT}} F_1(s). \qquad \square$$

Example 2.5. Find the Laplace transform of the *square-wave* function depicted in Figure 2.3. This bounded, piecewise continuous function is periodic of period $T = 2a$, and so its Laplace transform is given by

$$F(s) = \frac{1}{1 - e^{-2as}} F_1(s),$$

where

$$F_1(s) = \int_a^{2a} e^{-st}\, dt$$

$$= \frac{1}{s}(e^{-as} - e^{-2as}). \qquad (2.14)$$

Thus,

$$F(s) = \frac{e^{-as}}{s(1 + e^{-as})} = \frac{1}{s(1 + e^{as})}.$$

Observe that (2.13) can be written as

$$F(s) = \sum_{n=0}^{\infty} e^{-nTs} F_1(s) \qquad (x = \mathcal{Re}(s) > 0). \qquad (2.15)$$

In the case of the square-wave (Figure 2.3), the function can be expressed in the form

$$f(t) = u_a(t) - u_{2a}(t) + u_{3a}(t) - u_{4a}(t) + \cdots. \qquad (2.16)$$

Since $F_1(s) = (1/s)(e^{-as} - e^{-2as})$, we have from (2.15)

$$F(s) = \mathcal{L}(f(t)) = \sum_{n=0}^{\infty} e^{-2nas} \frac{1}{s}(e^{-as} - e^{-2as}) \qquad (T = 2a)$$

$$= \frac{1}{s} \sum_{n=0}^{\infty} (e^{-(2n+1)as} - e^{-(2n+2)as})$$

$$= \frac{1}{s}(e^{-as} - e^{-2as} + e^{-3as} - e^{-4as} + \cdots)$$

$$= \mathcal{L}(u_a(t)) - \mathcal{L}(u_{2a}(t)) + \mathcal{L}(u_{3a}(t)) - \mathcal{L}(u_{4a}(t)) + \cdots,$$

that is, we can take the Laplace transform of f term by term.

For other periodic functions with a representation as in (2.16), taking the Laplace transform in this fashion is often useful and justified.

Example 2.6.   The *half-wave-rectified* sine function is given by

$$f(t) = \begin{cases} \sin \omega t & \frac{2n\pi}{\omega} < t < \frac{(2n+1)\pi}{\omega} \\ 0 & \frac{(2n+1)\pi}{\omega} < t < \frac{(2n+2)\pi}{\omega}, \end{cases} \qquad n = 0, 1, 2, \ldots$$

(Figure 2.4). This bounded, piecewise continuous function is periodic with period $T = 2\pi/\omega$. Thus,

$$\mathcal{L}(f(t)) = \frac{1}{1 - e^{-\frac{2\pi s}{\omega}}} F_1(s),$$

where

$$F_1(s) = \int_0^{\frac{\pi}{\omega}} e^{-st} \sin \omega t \, dt$$

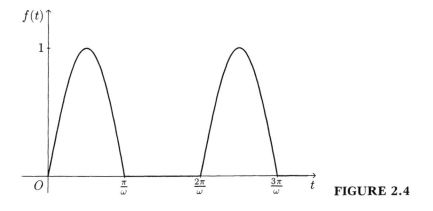

FIGURE 2.4

$$= \frac{e^{-st}}{s^2 + \omega^2}(-s \sin \omega t - \omega \cos \omega t)\Big|_0^{\frac{\pi}{\omega}}$$

$$= \frac{\omega}{s^2 + \omega^2}(1 + e^{-\frac{\pi s}{\omega}}).$$

Consequently,

$$\mathcal{L}(f(t)) = \frac{\omega}{(s^2 + \omega^2)(1 - e^{-\frac{\pi s}{\omega}})}.$$

The *full–wave–rectified sine* function (Figure 2.5)

$$f(t) = |\sin \omega t|,$$

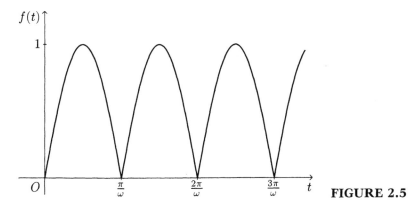

FIGURE 2.5

with $T = \pi/\omega$, has

$$\mathcal{L}\big(f(t)\big) = \frac{1}{1 - e^{-\frac{\pi s}{\omega}}} F_1(s)$$

$$= \frac{\omega}{s^2 + \omega^2} \left(\frac{1 + e^{-\frac{\pi s}{\omega}}}{1 - e^{-\frac{\pi s}{\omega}}} \right)$$

$$= \frac{\omega}{s^2 + \omega^2} \coth \frac{\pi s}{2\omega}.$$

Exercises 2.2

1. For Figures E.4–E.7, find the Laplace transform of the periodic function $f(t)$.

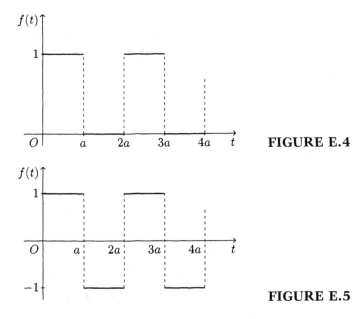

FIGURE E.4

FIGURE E.5

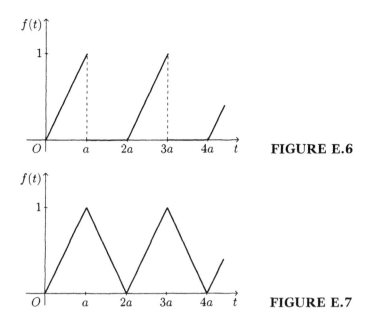

FIGURE E.6

FIGURE E.7

2. Compute the Laplace transform of the function

$$f(t) = u(t) - u_a(t) + u_{2a}(t) - u_{3a}(t) + \cdots$$

term by term and compare with Question 1(a).

3. Express the function in Question 1(b) as an infinite series of unit step functions and compute its Laplace transform term by term.

4. Determine $f(t) = \mathcal{L}^{-1}(F(s))$ for

$$F(s) = \frac{1 - e^{-as}}{s(e^{as} + e^{-as})} \qquad (\mathcal{R}e(s) > 0, \ a > 0)$$

by writing $F(s)$ as an infinite series of exponential functions and computing the inverse term by term. Draw a graph of $f(t)$ and verify that indeed $\mathcal{L}(f(t)) = F(s)$.

2.3 Derivatives

In order to solve differential equations, it is necessary to know the Laplace transform of the derivative f' of a function f. The virtue of $\mathcal{L}(f')$ is that it can be written in terms of $\mathcal{L}(f)$.

Theorem 2.7 (Derivative Theorem).

Suppose that f is continuous on $(0, \infty)$ and of exponential order α and that f' is piecewise continuous on $[0, \infty)$. Then

$$\mathcal{L}(f'(t)) = s\mathcal{L}(f(t)) - f(0^+) \qquad (\mathcal{R}e(s) > \alpha). \qquad (2.17)$$

PROOF. Integrating by parts,

$$\int_0^\infty e^{-st} f'(t)\, dt = \lim_{\substack{\delta \to 0 \\ \tau \to \infty}} \int_\delta^\tau e^{-st} f'(t)\, dt$$

$$= \lim_{\substack{\delta \to 0 \\ \tau \to \infty}} \left[e^{-st} f(t)\Big|_\delta^\tau + s \int_\delta^\tau e^{-st} f(t)\, dt \right]$$

$$= \lim_{\substack{\delta \to 0 \\ \tau \to \infty}} \left[e^{-s\tau} f(\tau) - e^{-s\delta} f(\delta) + s \int_\delta^\tau e^{-st} f(t)\, dt \right]$$

$$= -f(0^+) + s \int_0^\infty e^{-st} f(t)\, dt \qquad (\mathcal{R}e(s) > \alpha).$$

Therefore,

$$\mathcal{L}(f'(t)) = s\mathcal{L}(f(t)) - f(0^+).$$

We have made use of the fact that for $\mathcal{R}e(s) = x > \alpha$,

$$|e^{-s\tau} f(\tau)| \leq e^{-x\tau} M e^{\alpha\tau}$$

$$= M e^{-(x-\alpha)\tau} \to 0 \quad \text{as} \quad \tau \to \infty.$$

Also, note that $f(0^+)$ exists since $f'(0^+) = \lim_{t \to 0^+} f'(t)$ exists (see Exercises 2.3, Question 1). Clearly, if f is continuous at $t = 0$, then $f(0^+) = f(0)$ and our formula becomes

$$\mathcal{L}(f'(t)) = s\mathcal{L}(f(t)) - f(0). \qquad (2.18)$$

$\square$

Remark 2.8. An interesting feature of the derivative theorem is that we obtain $\mathcal{L}(f'(t))$ without requiring that f' itself be of exponential order. Example 1.14 was an example of this with $f(t) = \sin(e^{t^2})$.

Example 2.9. Let us compute $\mathcal{L}(\sin^2 \omega t)$ and $\mathcal{L}(\cos^2 \omega t)$ from (2.18). For $f(t) = \sin^2 \omega t$, we have $f'(t) = 2\omega \sin \omega t \cos \omega t = \omega \sin 2\omega t$. From

(2.18),

$$\mathcal{L}(\omega \sin 2\omega t) = s\,\mathcal{L}(\sin^2 \omega t) - \sin^2 0,$$

that is,

$$\mathcal{L}(\sin^2 \omega t) = \frac{1}{s}\,\mathcal{L}(\omega \sin 2\omega t)$$

$$= \frac{\omega}{s}\,\frac{2\omega}{s^2 + 4\omega^2}$$

$$= \frac{2\omega^2}{s(s^2 + 4\omega^2)}.$$

Similarly,

$$\mathcal{L}(\cos^2 \omega t) = \frac{1}{s}\,\mathcal{L}(-\omega \sin 2\omega t) + \frac{1}{s}$$

$$= -\frac{\omega}{s}\,\frac{2\omega}{s^2 + 4\omega^2} + \frac{1}{s}$$

$$= \frac{s^2 + 2\omega^2}{s(s^2 + 4\omega^2)}.$$

Note that if $f(0) = 0$, (2.18) can be expressed as

$$\mathcal{L}^{-1}\big(sF(s)\big) = f'(t),$$

where $F(s) = \mathcal{L}\big(f(t)\big)$. Thus, for example

$$\mathcal{L}^{-1}\left(\frac{s}{s^2 - a^2}\right) = \left(\frac{\sinh at}{a}\right)' = \cosh at.$$

It may be the case that f has a jump discontinuity other than at the origin. This can be treated in the following way.

Theorem 2.10. *Suppose that f is continuous on $[0, \infty)$ except for a jump discontinuity at $t = t_1 > 0$, and f has exponential order α with f' piecewise continuous on $[0, \infty)$. Then*

$$\mathcal{L}\big(f'(t)\big) = s\,\mathcal{L}\big(f(t)\big) - f(0) - e^{-t_1 s}\big(f(t_1^+) - f(t_1^-)\big) \qquad \big(\mathcal{R}e(s) > \alpha\big).$$

PROOF.

$$\int_0^\infty e^{-st} f'(t)\, dt$$

$$= \lim_{\tau \to \infty} \int_0^\tau e^{-st} f'(t)\, dt$$

$$= \lim_{\tau \to \infty} \left[e^{-st} f(t) \Big|_0^{t_1^-} + e^{-st} f(t) \Big|_{t_1^+}^\tau + s \int_0^\tau e^{-st} f(t)\, dt \right]$$

$$= \lim_{\tau \to \infty} \left[e^{-st_1} f(t_1^-) - f(0) + e^{-s\tau} f(\tau) - e^{-st_1} f(t_1^+) + s \int_0^\tau e^{-st} f(t)\, dt \right].$$

Hence

$$\mathcal{L}(f'(t)) = s\mathcal{L}(f(t)) - f(0) - e^{-st_1}\left(f(t_1^+) - f(t_1^-) \right).$$

If $0 = t_0 < t_1 < \cdots < t_n$ are a finite number of jump discontinuities, the formula becomes

$$\mathcal{L}(f'(t)) = s\mathcal{L}(f(t)) - f(0^+) - \sum_{k=1}^n e^{-st_k}\left(f(t_k^+) - f(t_k^-) \right). \qquad (2.19)$$

$\square$

Remark 2.11. If we assume that f' is continuous $[0, \infty)$ and also of exponential order, then it follows that the same is true of f itself.

To see this, suppose that

$$|f'(t)| \le M e^{\alpha t}, \qquad t \ge t_0, \ \alpha \ne 0.$$

Then

$$f(t) = \int_{t_0}^t f'(\tau)\, d\tau + f(t_0)$$

by the fundamental theorem of calculus, and

$$|f(t)| \le M \int_{t_0}^t e^{\alpha \tau}\, d\tau + |f(t_0)|$$

$$\le \frac{M}{\alpha} e^{\alpha t} + |f(t_0)|$$

$$\le C e^{\alpha t}, \qquad t \ge t_0.$$

Since f is continuous, the result holds for $\alpha \ne 0$, and the case $\alpha = 0$ is subsumed under this one.

To treat differential equations we will also need to know $\mathcal{L}(f'')$ and so forth. Suppose that for the moment we can apply formula (2.18) to f''. Then

$$\begin{aligned}
\mathcal{L}(f''(t)) &= s\mathcal{L}(f'(t)) - f'(0) \\
&= s(s\mathcal{L}(f(t)) - f(0)) - f'(0) \\
&= s^2\mathcal{L}(f(t)) - sf(0) - f'(0).
\end{aligned} \tag{2.20}$$

Similarly,

$$\begin{aligned}
\mathcal{L}(f'''(t)) &= s\mathcal{L}(f''(t)) - f''(0) \\
&= s^3\mathcal{L}(f(t)) - s^2f(0) - sf'(0) - f''(0)
\end{aligned} \tag{2.21}$$

under suitable conditions.

In the general case we have the following result.

Theorem 2.12. *Suppose that $f(t), f'(t), \cdots, f^{(n-1)}(t)$ are continuous on $(0, \infty)$ and of exponential order, while $f^{(n)}(t)$ is piecewise continuous on $[0, \infty)$. Then*

$$\mathcal{L}(f^{(n)}(t)) = s^n\mathcal{L}(f(t)) - s^{n-1}f(0^+) - s^{n-2}f'(0^+) - \cdots - f^{(n-1)}(0^+). \tag{2.22}$$

Example 2.13. Determine the Laplace transform of the Laguerre polynomials, defined by

$$L_n(t) = \frac{e^t}{n!}\frac{d^n}{dt^n}(t^n e^{-t}), \qquad n = 0, 1, 2, \ldots.$$

Let $y(t) = t^n e^{-t}$. Then

$$\mathcal{L}(L_n(t)) = \mathcal{L}\left(e^t \frac{1}{n!} y^{(n)}\right).$$

First, we find by Theorem 2.12, and subsequently the first translation theorem (1.27) coupled with (1.9),

$$\mathcal{L}(y^{(n)}) = s^n\mathcal{L}(y) = \frac{s^n n!}{(s+1)^{n+1}}.$$

It follows that

$$\mathcal{L}(L_n(t)) = \mathcal{L}\left(e^t \frac{1}{n!} y^{(n)}\right) = \frac{(s-1)^n}{s^{n+1}} \qquad (\mathcal{R}e(s) > 1),$$

again by the first translation theorem.

Exercises 2.3

1. In Theorem 2.7, prove that $f(0^+)$ exists?
(Hint: Consider for c sufficiently small,

$$\int_\delta^c f'(t)dt = f(c) - f(\delta),$$

and let $\delta \to 0^+$.)

2. Using the derivative theorem (2.7), show by mathematical induction that

$$\mathcal{L}(t^n) = \frac{n!}{s^{n+1}} \qquad (\mathcal{R}e(s) > 0), \; n = 1, 2, 3, \ldots.$$

3. (a) Show that

$$\mathcal{L}(\sinh \omega t) = \frac{\omega}{s^2 - \omega^2}$$

by letting $f(t) = \sinh \omega t$ and applying formula (2.20).

(b) Show that

$$\mathcal{L}(t \cosh \omega t) = \frac{s^2 + \omega^2}{(s^2 - \omega^2)^2}.$$

(c) Show that

$$\mathcal{L}(t \sinh \omega t) = \frac{2\omega s}{(s^2 - \omega^2)^2}.$$

4. Verify Theorem 2.10 for the function

$$f(t) = \begin{cases} t & 0 \le t \le 1 \\ 2 & t > 1. \end{cases}$$

5. Compute

(a) $\mathcal{L}(\sin^3 \omega t)$ \hspace{3cm} **(b)** $\mathcal{L}(\cos^3 \omega t)$.

6. Write out the details of the proof of Theorem 2.12.

7. Give an example to show that in Remark 2.11 the condition of continuity cannot be replaced by piecewise continuity.

2.4 Ordinary Differential Equations

The derivative theorem in the form of Theorem 2.12 opens up the possibility of utilizing the Laplace transform as a tool for solving ordinary differential equations. Numerous applications of the Laplace transform to ODEs will be found in ensuing sections.

Example 2.14. Consider the initial-value problem

$$\frac{d^2 y}{d t^2} + y = 1, \qquad y(0) = y'(0) = 0.$$

Let us assume for the moment that the solution $y = y(t)$ satisfies suitable conditions so that we may invoke (2.22). Taking the Laplace transform of both sides of the differential equation gives

$$\mathcal{L}(y'') + \mathcal{L}(y) = \mathcal{L}(1).$$

An application of (2.22) yields

$$s^2 \mathcal{L}(y) - s y(0) - y'(0) + \mathcal{L}(y) = \frac{1}{s},$$

that is,

$$\mathcal{L}(y) = \frac{1}{s(s^2 + 1)}.$$

Writing

$$\frac{1}{s(s^2 + 1)} = \frac{A}{s} + \frac{Bs + C}{s^2 + 1}$$

as a partial fraction decomposition, we find

$$\mathcal{L}(y) = \frac{1}{s} - \frac{s}{s^2 + 1}.$$

Applying the inverse transform gives the solution

$$y = 1 - \cos t.$$

One may readily check that this is indeed the solution to the initial-value problem.

Note that the initial conditions of the problem are absorbed into the method, unlike other approaches to problems of this type (i.e., the methods of *variation of parameters* or *undetermined coefficients*).

General Procedure. The Laplace transform method for solving ordinary differential equations can be summarized by the following steps.

(i) Take the Laplace transform of both sides of the equation. This results in what is called the *transformed equation*.

(ii) Obtain an equation $\mathcal{L}(y) = F(s)$, where $F(s)$ is an algebraic expression in the variable s.

(iii) Apply the inverse transform to yield the solution $y = \mathcal{L}^{-1}(F(s))$.

The various techniques for determining the inverse transform include partial fraction decomposition, translation, derivative and integral theorems, convolutions, and integration in the complex plane. All of these techniques except the latter are used in conjunction with standard tables of Laplace transforms.

Example 2.15. Solve

$$y''' + y'' = e^t + t + 1, \qquad y(0) = y'(0) = y''(0) = 0.$$

Taking $\mathcal{L}$ of both sides gives

$$\mathcal{L}(y''') + \mathcal{L}(y'') = \mathcal{L}(e^t) + \mathcal{L}(t) + \mathcal{L}(1),$$

or

$$[s^3 \mathcal{L}(y) - s^2 y(0) - s y'(0) - y''(0)]$$
$$+ [s^2 \mathcal{L}(y) - s y(0) - y'(0)] = \frac{1}{s-1} + \frac{1}{s^2} + \frac{1}{s}.$$

Putting in the initial conditions gives

$$s^3 \mathcal{L}(y) + s^2 \mathcal{L}(y) = \frac{2s^2 - 1}{s^2(s-1)},$$

which is

$$\mathcal{L}(y) = \frac{2s^2 - 1}{s^4(s+1)(s-1)}.$$

Applying a partial fraction decomposition to

$$\mathcal{L}(y) = \frac{2s^2 - 1}{s^4(s+1)(s-1)} = \frac{A}{s} + \frac{B}{s^2} + \frac{C}{s^3} + \frac{D}{s^4} + \frac{E}{s+1} + \frac{F}{s-1},$$

we find that

$$\mathcal{L}(y) = -\frac{1}{s^2} + \frac{1}{s^4} - \frac{1}{2(s+1)} + \frac{1}{2(s-1)},$$

and consequently

$$y = -\mathcal{L}^{-1}\left(\frac{1}{s^2}\right) + \mathcal{L}^{-1}\left(\frac{1}{s^4}\right) - \frac{1}{2}\,\mathcal{L}^{-1}\left(\frac{1}{s+1}\right) + \frac{1}{2}\,\mathcal{L}^{-1}\left(\frac{1}{s-1}\right)$$

$$= -t + \frac{1}{6}t^3 - \frac{1}{2}e^{-t} + \frac{1}{2}e^t.$$

In general, the Laplace transform method demonstrated above is particularly applicable to initial-value problems of nth-order linear ordinary differential equations with constant coefficients, that is,

$$a_n\frac{d^n y}{d\,t^n} + a_{n-1}\frac{d^{n-1}y}{d\,t^{n-1}} + \cdots + a_0 y = f(t),$$

$$(2.23)$$

$$y(0) = y_0, \quad y'(0) = y_1, \quad \ldots, \quad y^{(n-1)}(0) = y_{n-1}.$$

In engineering parlance, $f(t)$ is known as the *input, excitation,* or *forcing function,* and $y = y(t)$ is the *output* or *response.* In the event the input $f(t)$ has exponential order and be continuous, the output $y = y(t)$ to (2.23) can also be shown to have exponential order and be continuous (Theorem A.6). This fact helps to justify the application of the Laplace transform method (see the remark subsequent to Theorem A.6). More generally, when $f \in L$, the method can still be applied by assuming that the hypotheses of Theorem 2.12 are satisfied. While the solution $y = y(t)$ to (2.23) is given by the Laplace transform method for $t \geq 0$, it is in general valid on the whole real line, $-\infty < t < \infty$, if $f(t)$ has this domain.

Another important virtue of the Laplace transform method is that the input function $f(t)$ can be discontinuous.

Example 2.16. Solve

$$y'' + y = E\,u_a(t), \qquad y(0) = 0, \;\; y'(0) = 1.$$

Here the system is receiving an input of zero for $0 \leq t < a$ and E (constant) for $t \geq a$. Then

$$s^2\mathcal{L}(y) - sy(0) - y'(0) + \mathcal{L}(y) = \frac{E\,e^{-as}}{s}$$

and

$$\mathcal{L}(y) = \frac{1}{s^2 + 1} + \frac{E e^{-as}}{s(s^2 + 1)}$$

$$= \frac{1}{s^2 + 1} + E\left(\frac{1}{s} - \frac{s}{s^2 + 1}\right) e^{-as}.$$

Whence

$$y = \mathcal{L}^{-1}\left(\frac{1}{s^2 + 1}\right) + E\mathcal{L}^{-1}\left[\left(\frac{1}{s} - \frac{s}{s^2 + 1}\right) e^{-as}\right]$$

$$= \sin t + E u_a(t)(1 - \cos(t - a)),$$

by the second translation theorem (1.27). We can also express y in the form

$$y = \begin{cases} \sin t & 0 \leq t < a \\ \sin t + E(1 - \cos(t - a)) & t \geq a. \end{cases}$$

Note that $y(a^-) = y(a^+) = \sin a$, $y'(a^-) = y'(a^+) = \cos a$, $y''(a^-) = -\sin a$, but $y''(a^+) = -\sin a + E a^2$. Hence $y''(t)$ is only piecewise continuous.

Example 2.17. Solve

$$y'' + y = \begin{cases} \sin t & 0 \leq t \leq \pi \\ 0 & t > \pi \end{cases} \qquad y(0) = y'(0) = 0.$$

We have

$$s^2 \mathcal{L}(y) + \mathcal{L}(y) = \int_0^\pi e^{-st} \sin t \, dt$$

$$= \frac{-e^{-st}}{s^2 + 1} (s \cdot \sin t + \cos t)\bigg|_0^\pi$$

$$= \frac{e^{-\pi s}}{s^2 + 1} + \frac{1}{s^2 + 1}.$$

Therefore,

$$\mathcal{L}(y) = \frac{1}{(s^2 + 1)^2} + \frac{e^{-\pi s}}{(s^2 + 1)^2},$$

and by Example 2.42 (i) and the second translation theorem (1.31),

$$y = \frac{1}{2}(\sin t - t\cos t) + u_\pi(t)\left[\frac{1}{2}\big(\sin(t-\pi) - (t-\pi)\cos(t-\pi)\big)\right].$$

In other words,

$$y = \begin{cases} \frac{1}{2}(\sin t - t\cos t) & 0 \le t < \pi \\ -\frac{1}{2}\pi\cos t & t \ge \pi. \end{cases}$$

Observe that denoting the input function by $f(t)$,

$$f(t) = \sin t\big(1 - u_\pi(t)\big)$$
$$= \sin t + u_\pi(t)\sin(t - \pi),$$

from which

$$\mathcal{L}\big(f(t)\big) = \frac{1}{s^2 + 1} + \frac{e^{-\pi s}}{s^2 + 1},$$

again by the second translation theorem.

General Solutions. If the initial-value data of (2.23) are unspecified, the Laplace transform can still be applied in order to determine the general solution.

Example 2.18. Consider

$$y'' + y = e^{-t},$$

and let $y(0) = y_0$, $y'(0) = y_1$ be unspecified. Then

$$s^2\mathcal{L}(y) - s\,y(0) - y'(0) + \mathcal{L}(y) = \mathcal{L}(e^{-t}),$$

that is,

$$\mathcal{L}(y) = \frac{1}{(s+1)(s^2+1)} + \frac{s\,y_0}{s^2+1} + \frac{y_1}{s^2+1}$$

$$= \frac{\frac{1}{2}}{s+1} - \frac{\frac{1}{2}s - \frac{1}{2}}{s^2+1} + \frac{y_0 s}{s^2+1} + \frac{y_1}{s^2+1},$$

by taking a partial fraction decomposition. Applying $\mathcal{L}^{-1}$,

$$y = \frac{1}{2}e^{-t} + \left(y_0 - \frac{1}{2}\right)\cos t + \left(y_1 + \frac{1}{2}\right)\sin t.$$

Since y_0, y_1 can take on all possible values, the general solution to the problem is given by

$$y = c_0 \cos t + c_1 \sin t + \tfrac{1}{2} e^{-t},$$

where c_0, c_1 are arbitrary real constants. Note that this solution is valid for $-\infty < t < \infty$.

Boundary-Value Problems. This type of problem is also amenable to solution by the Laplace transform method. As a typical example consider

$$y'' + \lambda^2 y = \cos \lambda t, \qquad y(0) = 1, \quad y\left(\frac{\pi}{2\lambda}\right) = 1.$$

Then

$$\mathcal{L}(y'') + \lambda^2 \mathcal{L}(y) = \mathcal{L}(\cos \lambda t),$$

so that

$$(s^2 + \lambda^2)\mathcal{L}(y) = \frac{s}{s^2 + \lambda^2} + s y(0) + y'(0),$$

implying

$$\mathcal{L}(y) = \frac{s}{(s^2 + \lambda^2)^2} + \frac{s y(0)}{s^2 + \lambda^2} + \frac{y'(0)}{s^2 + \lambda^2}.$$

Therefore,

$$y = \frac{1}{2\lambda} t \sin \lambda t + \cos \lambda t + \frac{y'(0)}{\lambda} \sin \lambda t, \qquad (2.24)$$

where we have invoked Example 2.42 (ii) to determine the first term and replaced $y(0)$ with its value of 1. Finally, from (2.24)

$$1 = y\left(\frac{\pi}{2\lambda}\right) = \frac{\pi}{4\lambda^2} + \frac{y'(0)}{\lambda}$$

gives

$$\frac{y'(0)}{\lambda} = 1 - \frac{\pi}{4\lambda^2},$$

and thus

$$y = \frac{1}{2\lambda} t \sin \lambda t + \cos \lambda t + \left(1 - \frac{\pi}{4\lambda^2}\right) \sin \lambda t.$$

Similarly, if the boundary data had been, say

$$y(0) = 1, \qquad y'\left(\frac{\pi}{\lambda}\right) = 1,$$

then differentiating in (2.24)

$$y' = \frac{1}{2\lambda}(\sin \lambda t + \lambda t \cos \lambda t) - \lambda \sin \lambda t + y'(0) \cos \lambda t.$$

Thus,

$$1 = y'\left(\frac{\pi}{\lambda}\right) = \frac{-\pi}{2\lambda} - y'(0)$$

and

$$y'(0) = -\left(1 + \frac{\pi}{2\lambda}\right),$$

to yield

$$y = \frac{1}{2\lambda} t \sin \lambda t + \cos \lambda t - \frac{1}{\lambda}\left(1 + \frac{\pi}{2\lambda}\right) \sin \lambda t.$$

Systems of Differential Equations. Systems of differential equations can also be readily handled by the Laplace transform method. We illustrate with a few examples.

Example 2.19.

$$\frac{dy}{dt} = -z; \quad \frac{dz}{dt} = y, \quad y(0) = 1, \quad z(0) = 0.$$

Then

$$\mathcal{L}(y') = -\mathcal{L}(z) \qquad \text{i.e., } s\mathcal{L}(y) - 1 = -\mathcal{L}(z)$$

and $\hspace{6cm}$ (2.25)

$$\mathcal{L}(z') = \mathcal{L}(y) \qquad \text{i.e., } s\mathcal{L}(z) = \mathcal{L}(y).$$

Solving the simultaneous equation (2.25)

$$s^2 \mathcal{L}(y) - s = -s\mathcal{L}(z) = -\mathcal{L}(y),$$

or

$$\mathcal{L}(y) = \frac{s}{s^2 + 1},$$

so that

$$y = \cos t, \qquad z = -y' = \sin t.$$

Example 2.20.

$$y' + z' + y + z = 1,$$
$$y' + z = e^t, \quad y(0) = -1, \ z(0) = 2.$$

From the first equation, we have

$$s\mathcal{L}(y) + 1 + s\mathcal{L}(z) - 2 + \mathcal{L}(y) + \mathcal{L}(z) = \frac{1}{s}. \qquad (2.26)$$

From the second equation, we have

$$s\mathcal{L}(y) + 1 + \mathcal{L}(z) = \frac{1}{s-1}. \qquad (2.27)$$

Solving (2.26) and (2.27), we arrive at

$$\mathcal{L}(y) = \frac{-s^2 + s + 1}{s(s-1)^2}$$

$$= \frac{1}{s} - \frac{2}{s-1} + \frac{1}{(s-1)^2}.$$

Taking the inverse transform yields

$$y = 1 - 2e^t + t\,e^t, \qquad z = 2e^t - t\,e^t.$$

Integrals. In certain differential equations it is also necessary to compute the Laplace transform of an integral.

Theorem 2.21. *If f is piecewise continuous on $[0, \infty)$ of exponential order $\alpha(\geq 0)$, and*

$$g(t) = \int_0^t f(u)\,du,$$

then

$$\mathcal{L}(g(t)) = \frac{1}{s}\mathcal{L}(f(t)) \qquad (\mathcal{R}e(s) > \alpha).$$

PROOF. Since $g'(t) = f(t)$ except at points of discontinuity of f, integration by parts gives

$$\int_0^\infty e^{-st}g(t)\,dt = \lim_{\tau \to \infty}\left[\frac{g(t)e^{-st}}{-s}\bigg|_0^\tau + \frac{1}{s}\int_0^\tau e^{-st}f(t)\,dt\right].$$

Since $g(0) = 0$, we need only compute

$$\lim_{\tau \to \infty} \frac{g(\tau)e^{-s\tau}}{-s}.$$

To this end,

$$|g(\tau)e^{-s\tau}| \leq e^{-x\tau} \int_0^\tau |f(u)|\, du$$

$$\leq M e^{-x\tau} \int_0^\tau e^{\alpha u}\, du$$

$$= \frac{M}{\alpha}(e^{-(x-\alpha)\tau} - e^{-x\tau})$$

$$\to 0 \quad \text{as} \quad \tau \to \infty \quad \text{for} \quad x = \mathcal{R}e(s) > \alpha > 0.$$

Similarly, this holds for $\alpha = 0$. Hence

$$\mathcal{L}(g(t)) = \frac{1}{s}\mathcal{L}(f(t)) \qquad (\mathcal{R}e(s) > \alpha). \qquad \qquad \square$$

Example 2.22.

$$\mathcal{L}(\mathrm{Si}(t)) = \mathcal{L}\left(\int_0^t \frac{\sin u}{u}\, du\right) = \frac{1}{s}\mathcal{L}\left(\frac{\sin t}{t}\right)$$

$$= \frac{1}{s}\tan^{-1}\frac{1}{s},$$

by Example 1.34 (i). The function, $\mathrm{Si}(t)$, is called the *sine integral*.

The result of Theorem 2.22 can also be expressed in the form

$$\mathcal{L}^{-1}\left(\frac{F(s)}{s}\right) = \int_0^t f(u)\, du,$$

where $F(s) = \mathcal{L}(f(t))$. Hence, for example,

$$\mathcal{L}^{-1}\left(\frac{1}{s(s^2 - a^2)}\right) = \frac{1}{a}\int_0^t \sinh au\, du = \frac{1}{a^2}(\cosh at - 1).$$

Differential equations that involve integrals (known as *integro-differential* equations) commonly arise in problems involving electrical circuits.

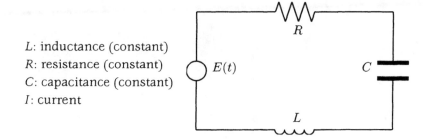

FIGURE 2.6

Electrical Circuits. In the (*RCL*) circuit in Figure 2.6, the voltage drops across the inductor, resistor, and capacitor are given by $L(dI/dt)$, RI, and $(1/C)\int_0^t I(\tau)\,d\tau$, respectively, where Kirchoff's voltage law states that *the sum of the voltage drops across the individual components equals the impressed voltage*, $E(t)$, that is,

$$L\frac{dI}{dt} + RI + \frac{1}{C}\int_0^t I(\tau)\,d\tau = E(t). \tag{2.28}$$

Setting $Q(t) = \int_0^t I(\tau)\,d\tau$ (the charge of the condenser), we can write (2.28) as

$$L\frac{d^2Q}{dt^2} + R\frac{dQ}{dt} + \frac{Q}{C} = E(t) \tag{2.29}$$

since $I = dQ/dt$. This will be the basis of some of the electrical circuit problems throughout the sequel.

Example 2.23. Suppose that the current I in an electrical circuit satisfies

$$L\frac{dI}{dt} + RI = E_0 \sin \omega t,$$

where L, R, E_0, and ω are constants. Find $I = I(t)$ for $t > 0$ if $I(0) = 0$.
  Taking the Laplace transform,

$$Ls\,\mathcal{L}(I) + R\,\mathcal{L}(I) = \frac{E_0\omega}{s^2 + \omega^2},$$

that is,

$$\mathcal{L}(I) = \frac{E_0\omega}{(Ls + R)(s^2 + \omega^2)}.$$

Considering partial fractions

$$\mathcal{L}(I) = \frac{E_0\omega/L}{(s+R/L)(s^2+\omega^2)} = \frac{A}{s+R/L} + \frac{Bs+C}{s^2+\omega^2},$$

we find that

$$A = \frac{E_0L\omega}{L^2\omega^2+R^2}, \quad B = \frac{-E_0L\omega}{L^2\omega^2+R^2}, \quad C = \frac{E_0R\omega}{L^2\omega^2+R^2},$$

and so

$$I(t) = \frac{E_0L\omega}{L^2\omega^2+R^2}\,e^{-\frac{R}{L}t} + \frac{E_0R}{L^2\omega^2+R^2}\sin\omega t - \frac{E_0L\omega}{L^2\omega^2+R^2}\cos\omega t.$$

Example 2.24. Suppose that the current I in the electrical circuit depicted in Figure 2.7 satisfies

$$L\frac{dI}{dt} + \frac{1}{C}\int_0^t I(\tau)\,d\tau = E,$$

where L, C, and E are positive constants, $I(0) = 0$. Then

$$Ls\,\mathcal{L}(I) + \frac{\mathcal{L}(I)}{Cs} = \frac{E}{s},$$

implying

$$\mathcal{L}(I) = \frac{EC}{LCs^2+1} = \frac{E}{L(s^2+1/LC)}.$$

Thus,

$$I(t) = E\sqrt{\frac{C}{L}}\,\sin\frac{1}{\sqrt{LC}}t.$$

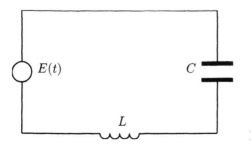

FIGURE 2.7

Differential Equations with Polynomial Coefficients. Recall (Theorem 1.34) that for $F(s) = \mathcal{L}(y(t))$,

$$\frac{d^n}{ds^n} F(s) = (-1)^n \mathcal{L}(t^n y(t)) \qquad (s > \alpha)$$

for $y(t)$ piecewise continuous on $[0, \infty)$ and of exponential order α. Hence, for $n = 1$,

$$\mathcal{L}(ty(t)) = -F'(s).$$

Suppose further that $y'(t)$ satisfies the hypotheses of the theorem. Then

$$\mathcal{L}(ty'(t)) = -\frac{d}{ds} \mathcal{L}(y'(t))$$

$$= -\frac{d}{ds} (sF(s) - y(0))$$

$$= -sF'(s) - F(s).$$

Similarly, for $y''(t)$,

$$\mathcal{L}(ty'') = -\frac{d}{ds} \mathcal{L}(y''(t))$$

$$= -\frac{d}{ds} (s^2 F(s) - sy(0) - y'(0))$$

$$= -s^2 F'(s) - 2sF(s) + y(0).$$

In many cases these formulas for $\mathcal{L}(ty(t))$, $\mathcal{L}(ty'(t))$, and $\mathcal{L}(ty''(t))$ can be used to solve linear differential equations whose coefficients are (first-degree) polynomials.

Example 2.25. Solve

$$y'' + ty' - 2y = 4, \quad y(0) = -1, \quad y'(0) = 0.$$

Taking the Laplace transform of both sides yields

$$s^2 F(s) + s - (sF'(s) + F(s)) - 2F(s) = \frac{4}{s},$$

or

$$F'(s) + \left(\frac{3}{s} - s\right) F(s) = -\frac{4}{s^2} + 1.$$

The integrating factor is

$$\mu(s) = e^{\int (\frac{3}{s} - s) ds} = s^3 e^{-s^2/2}.$$

Therefore,

$$\left(F(s) s^3 e^{-s^2/2} \right)' = -\frac{4}{s^2} s^3 e^{-s^2/2} + s^3 e^{-s^2/2},$$

and

$$F(s) s^3 e^{-s^2/2} = -4 \int s e^{-s^2/2} ds + \int s^3 e^{-s^2/2} ds.$$

Substituting $u = -s^2/2$ into both integrals gives

$$F(s) s^3 e^{-s^2/2} = 4 \int e^u du + 2 \int u e^u du$$

$$= 4 e^{-s^2/2} + 2 \left(\frac{-s^2}{2} e^{-s^2/2} - e^{-s^2/2} \right) + C$$

$$= 2 e^{-s^2/2} - s^2 e^{-s^2/2} + C.$$

Thus,

$$F(s) = \frac{2}{s^3} - \frac{1}{s} + \frac{C}{s^3} e^{s^2/2}.$$

Since $F(s) \to 0$ as $s \to \infty$, we must have $C = 0$ and

$$y(t) = t^2 - 1,$$

which can be verified to be the solution.

There are pitfalls, however, of which the reader should be aware. A seemingly innocuous problem such as

$$y' - 2ty = 0, \qquad y(0) = 1,$$

has $y(t) = e^{t^2}$ as its solution, and this function, as we know, does not possess a Laplace transform. (See what happens when you try to apply the Laplace transform method to this problem.)

Another caveat is that if the differential equation has a regular singular point, one of the solutions may behave like $\log t$ as $t \to 0^+$; hence its derivative has no Laplace transform (see Exercises 1.2, Question 3). In this case, the Laplace transform method can deliver only the solution that is bounded at the origin.

Example 2.26. Solve

$$ty'' + y' + 2y = 0.$$

The point $t = 0$ is a regular singular point of the equation. Let us determine the solution that satisfies $y(0) = 1$. Taking the Laplace transform,

$$\left(-s^2 F'(s) - 2sF(s) + 1 \right) + \left(sF(s) - 1 \right) + 2F(s) = 0,$$

that is,

$$-s^2 F'(s) - sF(s) + 2F(s) = 0,$$

or

$$F'(s) + \left(\frac{1}{s} - \frac{2}{s^2} \right) F(s) = 0, \qquad s > 0.$$

Then the integrating factor is

$$\mu(s) = e^{\int \left(\frac{1}{s} - \frac{2}{s^2} \right) ds} = se^{2/s}.$$

Therefore,

$$\left(F(s)se^{2/s} \right)' = 0$$

and

$$F(s) = \frac{Ce^{-2/s}}{s}.$$

Taking the series $e^x = \sum_{n=0}^{\infty} (x^n/n!)$ with $x = -2/s$ implies

$$F(s) = C \sum_{n=0}^{\infty} \frac{(-1)^n 2^n}{n! s^{n+1}}.$$

In view of (2.8) we can take $\mathcal{L}^{-1}$ term by term so that

$$y(t) = C \sum_{n=0}^{\infty} \frac{(-1)^n 2^n t^n}{(n!)^2}.$$

The condition $y(0) = 1$ gives $C = 1$.

Note that $y(t) = J_0(2\sqrt{at})$ with $a = 2$, from the table of Laplace transforms (pp. 210–218), where J_0 is the well-known Bessel function (2.55). There is another solution to this differential equation which is unbounded at the origin and cannot be determined by the preceding method.

Exercises 2.4

1. Solve the following initial-value problems by the Laplace transform method.

   (a) $\dfrac{dy}{dt} - y = \cos t$, $y(0) = -1$

   (b) $\dfrac{dy}{dt} + y = t^2 e^t$, $y(0) = 2$

   (c) $\dfrac{d^2 y}{dt^2} + 4y = \sin t$, $y(0) = 1$, $y'(0) = 0$

   (d) $\dfrac{d^2 y}{dt^2} - 2\dfrac{dy}{dt} - 3y = te^t$, $y(0) = 2$, $y'(0) = 1$

   (e) $\dfrac{d^3 y}{dt^3} + 5\dfrac{d^2 y}{dt^2} + 2\dfrac{dy}{dt} - 8y = \sin t$, $y(0) = 0$, $y'(0) = 0$,
       $y''(0) = -1$

   (f) $\dfrac{d^2 y}{dt^2} + \dfrac{dy}{dt} = f(t)$, $y(0) = 1$, $y'(0) = -1$, where
       $f(t) = \begin{cases} 1 & 0 < t < 1 \\ 0 & t > 1 \end{cases}$

   (g) $y'' + y = \begin{cases} \cos t & 0 \le t \le \pi \\ 0 & t > \pi, \end{cases}$ $y(0) = 0, y'(0) = 0$

   (h) $y^{(4)} - y = 0$, $y(0) = 1$, $y'(0) = y''(0) = y'''(0) = 0$.

2. Solve the boundary value problems.

   (a) $y'' + \lambda^2 y = \sin \lambda t$, $y(0) = 1$, $y\left(\frac{\pi}{2\lambda}\right) = \pi$.

   (b) $y'' + \lambda^2 y = t$, $y(0) = 1$, $y'\left(\frac{\pi}{\lambda}\right) = -1$.

3. Suppose that the current I in an electrical circuit satisfies

   $$L\frac{dI}{dt} + RI = E_0,$$

   where L, R, E_0 are positive constants.

   (a) Find $I(t)$ for $t > 0$ if $I(0) = I_0 > 0$.
   (b) Sketch a graph of the solution in (a) for the case $I_0 > E_0/R$.
   (c) Show that $I(t)$ tends to E_0/R as $t \to \infty$.

4. Suppose that the current I in an electrical circuit satisfies

   $$L\frac{dI}{dt} + RI = E_0 + A\cos\omega t,$$

where L, R, E_0, A and ω are constants. Find $I(t)$ for $t > 0$ if $I(0) = 0$.

5. Find the current $I(t)$, $t > 0$, if

$$L\frac{dI}{dt} + RI + \frac{1}{C}\int_0^t I(\tau)\,d\tau = \sin t,$$

and $L = 1$, $R = 3$, $C = \frac{1}{2}$, $I(0) = 1$.

6. Solve the following systems of equations by the Laplace transform method.

(a) $2\dfrac{dx}{dt} + 3x + y = 0$

   $2\dfrac{dy}{dt} + x + 3y = 0$

   $x(0) = 2, \quad y(0) = 0$

(b) $\dfrac{dx}{dt} + x - y = 1 + \sin t$

   $\dfrac{dy}{dt} - \dfrac{dx}{dt} + y = t - \sin t$

   $x(0) = 0, \quad y(0) = 1$

(c) $x(t) - y''(t) + y(t) = e^{-t} - 1$

   $x'(t) + y'(t) - y(t) = -3e^{-t} + t$

   $x(0) = 0, \quad y(0) = 1, \quad y'(0) = -2.$

7. Solve the following differential equations by the Laplace transform method.

   (a) $ty' - y = 1$
   (b) $ty'' - y' = -1, \quad y(0) = 0$
   (c) $ty'' + y = 0, \quad y(0) = 0$
   (d) $ty'' + (t+1)y' + 2y = e^{-t}, \quad y(0) = 0.$

2.5 Dirac Operator

In order to model certain physical events mathematically, such as a sudden power surge caused by a bolt of lightning, or the dynamical effects of a hammer blow to a spring-mounted weight, it turns out that ordinary functions are ill suited for these purposes. What is required is an entirely new entity that is not really a function at all.

Because of the status of this new entity, we also require a new tool in order to discuss it, namely the Riemann–Stieltjes integral, which is just a natural extension of the conventional Riemann integral.

Riemann–Stieltjes Integral. Consider a partition of the interval $[\alpha, \beta]$ given by $\alpha = t_0 < t_1 < \cdots < t_{n-1} < t_n = \beta$, choosing from each subinterval $[t_{i-1}, t_i]$ an arbitrary point x_i with $t_{i-1} \le x_i \le t_i$. Given functions f and φ defined on $[\alpha, \beta]$, we form the sum

$$\sum_{i=1}^{n} f(x_i)[\varphi(t_i) - \varphi(t_{i-1})]. \tag{2.30}$$

If these sums converge to a finite limit L as $\Delta = \max_i(t_i - t_{i-1}) \to 0$ as $i \to \infty$, and for every choice of $x_i \in [t_{i-1}, t_i]$, then this limit is called the *Riemann-Stieltjes integral of f with respect to φ on $[\alpha, \beta]$*, and for the value of L we write

$$\int_{\alpha}^{\beta} f(t) \, d\varphi(t).$$

If $\varphi(t) = t$, then all the sums in (2.30) are the usual Riemann sums and we obtain the ordinary Riemann integral.

The basic properties of the Riemann–Stieltjes integral are listed in the Appendix (Theorem A.9) and are very similar to those of the Riemann integral, as one might expect. It is important to note that the function φ need not be continuous. In fact, *if f is continuous on $[\alpha, \beta]$ and φ is a nondecreasing function on $[\alpha, \beta]$, then $\int_{\alpha}^{\beta} f(t) \, d\varphi(t)$ exists* (see Protter and Morrey [10], Theorem 12.16).

For example, let $\varphi(t) = u_a(t)$, the unit step function (Example 1.25):

$$\varphi(t) = \begin{cases} 1 & t \ge a \\ 0 & t < a, \end{cases}$$

for $a \ge 0$. If f is continuous on some interval $[\alpha, \beta]$ containing a, say $\alpha < a < \beta$, then for any particular partition, only for $t_{j-1} < a \le t_j$ is there any contribution to the integral (all the other terms being zero), and

$$\sum_{i=1}^{n} f(x_i)[\varphi(t_i) - \varphi(t_{i-1})] = f(x_j)[\varphi(t_j) - \varphi(t_{j-1})]$$

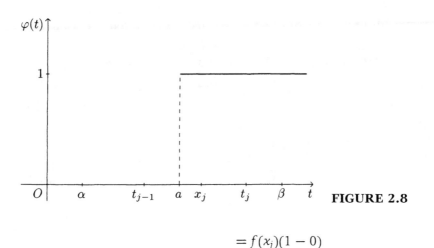

FIGURE 2.8

$$= f(x_j)(1 - 0)$$

$$= f(x_j)$$

(Figure 2.8). Taking the limit as $\Delta \to 0$ (whereby $x_j \to a$) gives the value of the Riemann–Stieltjes integral

$$\int_\alpha^\beta f(t)\, du_a(t) = f(a) \tag{2.31}$$

since $f(x_j) \to f(a)$.

The property (2.31) is called the "sifting property" in that it sifts out one particular value of the function f. Let us denote

$$\delta_a = du_a, \qquad a \geq 0, \tag{2.32}$$

and for $a = 0$, set $\delta = \delta_0$. From the sifting property we see that the action of δ_a on continuous functions is that of an *operator*, that is,

$$\delta_a[f] = \int_{-\infty}^\infty f(t)\, \delta_a(t) = f(a), \tag{2.33}$$

and we see that this operator is *linear*:

$$\delta_a[c_1 f_1 + c_2 f_2] = c_1 \delta_a[f_1] + c_2 \delta_a[f_2],$$

for constants c_1, c_2.

We shall call δ_a the *Dirac operator*, although it is also known as the *Dirac distribution, Dirac measure concentrated at* a, *Dirac delta function,* or *impulse function.* P.A.M. Dirac, one of the founders of quantum mechanics, made extensive use of it in his work. However, the "delta function" was highly controversial until it was made rigorous by the French mathematician Laurent Schwartz, in his book *Théorie des distributions* (cf. [13]). The class of linear operators of which the Dirac operator is just one example is known as *distributions* or *generalized functions* (see Guest [5], Chapter 12; also Richards and Youn [11]).

Let us use the Riemann integral to show that the sifting property (2.33) for continuous functions could not possibly hold for any "proper function" φ_a.

Let f_n be continuous, $f_n(t) = 0$ for $|t| \geq 1/n$, $f_n(t) = 1 - n|t|$ for $|t| < 1/n$, so that with $a = 0$, $f_n(0) = 1$. If φ_0 is Riemann integrable, then it must be bounded by some constant M on, say, $[-1, 1]$. If φ_0 satisfies the sifting property, it follows that

$$1 = \int_{-\infty}^{\infty} f_n(t)\,\varphi_0(t)\,dt \leq \int_{-\frac{1}{n}}^{\frac{1}{n}} f_n(t)|\varphi_0(t)|\,dt$$

$$\leq M \int_{-\frac{1}{n}}^{\frac{1}{n}} dt = \frac{2M}{n},$$

a contradiction for n sufficiently large (Figure 2.9).

However, there is an important relationship between the Riemann–Stieltjes and Riemann integrals under suitable conditions. Notably, *if f, φ, φ' are continuous on $[\alpha, \beta]$, then*

$$\int_{\alpha}^{\beta} f(t)\,d\varphi(t) = \int_{\alpha}^{\beta} f(t)\,\varphi'(t)\,dt \qquad (2.34)$$

(see Theorem A.10).

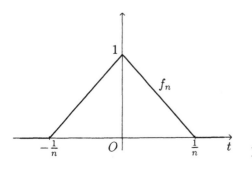

FIGURE 2.9

One further property of the Dirac operator worth noting is

$$\int_{-\infty}^{\infty} \delta_a(t) = 1, \tag{2.35}$$

which can be expressed as *the total point mass concentrated at a is unity.*

Laplace Transform. In terms of the Riemann–Stieltjes integral, the *Laplace transform with respect to a function* φ defined on $[0, \infty)$ is given by [cf. Widder [15] for an explication using this approach]

$$F(s) = \mathcal{L}_{R-S}(d\varphi) = \int_{-\infty}^{\infty} e^{-st} d\varphi(t) = \lim_{b \to \infty} \int_{-b}^{b} e^{-st} d\varphi(t) \tag{2.36}$$

whenever this integral exists. Since we have taken the integral over $(-\infty, \infty)$, we will always set $\varphi(t) = 0$ for $t < 0$. In particular, for $d\varphi = du_a = \delta_a$,

$$\mathcal{L}_{R-S}(\delta_a) = \int_{-\infty}^{\infty} e^{-st} \delta_a(t)$$

$$= e^{-as}, \qquad a \geq 0, \tag{2.37}$$

by the sifting property. When $a = 0$, we have

$$\mathcal{L}_{R-S}(\delta) = 1.$$

Here we have an instance of the basic property of the Laplace transform, $F(s) \to 0$ as $s \to \infty$, being violated. But of course, δ is not a function but a linear operator.

The application of the Riemann–Stieltjes Laplace transform (or *Laplace-Stieltjes transform* as it is known) becomes more transparent with the following result. We will take a slight liberty here with the notation and write $\mathcal{L}_{R-S}(\psi')$ for $\mathcal{L}_{R-S}(d\psi)$ whenever ψ' is continuous on $[0, \infty)$.

Theorem 2.27. *Suppose that φ is a continuous function of exponential order on $[0, \infty)$. Then*

$$\mathcal{L}_{R-S}(\varphi) = \mathcal{L}(\varphi).$$

PROOF. Let

$$\psi(t) = \int_0^t \varphi(\tau) \, d\tau,$$

and set $\varphi(t) = \psi(t) = 0$ for $t < 0$. Then $\psi'(t) = \varphi(t)$, except possibly at $t = 0$, and in view of (2.34),

$$\mathcal{L}(\varphi) = \int_{-\infty}^{\infty} e^{-st}\varphi(t)\,dt = \int_{-\infty}^{\infty} e^{-st}\psi'(t)\,dt = \int_{-\infty}^{\infty} e^{-st}d\psi(t)$$

$$= \mathcal{L}_{R-S}(d\psi) = \mathcal{L}_{R-S}(\psi') = \mathcal{L}_{R-S}(\varphi),$$

as desired. □

Remark 2.28. In the preceding theorem, the continuous function φ need not be of exponential order as long as the usual Laplace transform exists.

Thus we have the following general principle:
When taking the Laplace–Stieltjes transform $\mathcal{L}_{R-S}$ of functions in a differential equation, we may instead take the ordinary Laplace transform, $\mathcal{L}$.

Example 2.29. Let us solve the differential equation

$$x'(t) = \delta(t), \qquad x(0) = 0.$$

First note that this equation can be interpreted in the sense that both sides are linear operators $\left[\text{i.e., } x'[f] = \int_{-\infty}^{\infty} f(t)x'(t)\,dt \text{ for, say,}\right.$ continuous f, which vanishes outside a finite interval$\left.\right]$. Applying $\mathcal{L}_{R-S}$ to both sides,

$$s\mathcal{L}(x) = \mathcal{L}_{R-S}(\delta) = 1,$$

and

$$\mathcal{L}(x) = \frac{1}{s},$$

so that $x(t) \equiv 1$, $t \geq 0$.

Note, however, that the initial condition $x(0) = 0$ is not satisfied, but if we define $x(t) = 0$ for $t < 0$, then $\lim_{t \to 0^-} x(t) = 0$. Moreover,

$$x(t) = u(t),$$

the unit step function (compare with (2.32)).

Applications. The manner in which the Dirac operator has come to be used in modeling a sudden impulse comes from consideration

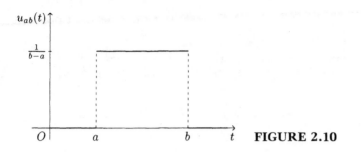

FIGURE 2.10

of the step function (Example 1.26):

$$u_{ab}(t) = \frac{1}{b-a}\left(u_a(t) - u_b(t)\right), \qquad b > a \geq 0$$

(Figure 2.10). Note that $u_{ab}(t)$ has the property

$$\int_{-\infty}^{\infty} u_{ab}(t)\,dt = 1. \qquad (2.38)$$

In order to simulate a sudden "impulse," we let b approach a and define

$$\Delta_a(t) = \lim_{b\to a} u_{ab}(t). \qquad (2.39)$$

Then $\Delta_a(t) = 0$ for all $t \neq a$ and is undefined (or ∞ if you like) at $t = a$.

From another perspective, let f be continuous in some interval $[\alpha, \beta]$ containing a, with $\alpha < a < b < \beta$. Then

$$\int_{-\infty}^{\infty} f(t)\,u_{ab}(t)\,dt = \frac{1}{b-a}\int_a^b f(t)\,dt$$

$$= f(c)$$

for some point $c \in [a, b]$ by the mean-value theorem for integrals (Figure 2.11). Taking the limit as $b \to a$, we get $f(c) \to f(a)$, that is,

$$\lim_{b\to a} \int_{-\infty}^{\infty} f(t)\,u_{ab}(t)\,dt = f(a). \qquad (2.40)$$

This suggests in a heuristic sense that *if only* we could take this limit inside the integral (which is not possible), then coupled with

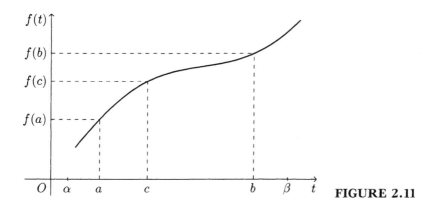

FIGURE 2.11

(2.39) we would arrive at the expression

$$\int_{-\infty}^{\infty} f(t) \, \Delta_a(t) \, dt = f(a).$$

This formula has absolutely no meaning in the Riemann integral sense (remember that $\Delta_a(t)$ is zero except for the value ∞ at $t = a$), but we have already given such an expression a rigorous meaning in the Riemann–Stieltjes sense of (2.33).

Again, in (2.38), if only one could take the limit as $b \to a$ inside the integral, we would have

$$\int_{-\infty}^{\infty} \Delta_a(t) \, dt = 1,$$

also achieved rigorously in (2.35).

As far as the Laplace transform goes, we have

$$\mathcal{L}(u_{ab}(t)) = \frac{e^{-as} - e^{-bs}}{s(b-a)}$$

as in Example 1.26. Letting $b \to a$ and applying l'Hôpital's rule,

$$\lim_{b \to a} \mathcal{L}(u_{ab}(t)) = \lim_{b \to a} \frac{e^{-as} - e^{-bs}}{s(b-a)} = e^{-as}. \tag{2.41}$$

Since $\lim_{b \to a} u_{ab}(t) = \Delta_a(t)$, it is tempting (but meaningless) to write

$$\lim_{b \to a} \mathcal{L}(u_{ab}(t)) = \mathcal{L}(\Delta_a(t)),$$

and hence by (2.41), equating the two limits

$$\mathcal{L}\big(\Delta_a(t)\big) = e^{-as}.$$

This is just the expression obtained in (2.37).

The foregoing illustrates that the mathematical modeling of a sudden impulse is achieved rigorously by the treatment given in terms of the Riemann–Stieltjes integral.

Hereafter, for the sake of convenience, we will abuse the notation further and simply write

$$\mathcal{L}(\delta_a) = e^{-as}.$$

Example 2.30. A pellet of mass m is fired from a gun at time $t = 0$ with a muzzle velocity v_0. If the pellet is fired into a viscous gas, the equation of motion can be expressed as

$$m\frac{d^2x}{dt^2} + k\frac{dx}{dt} = m\,v_0\,\delta(t), \qquad x(0) = 0, \;\; x'(0) = 0,$$

where $x(t)$ is the displacement at time $t \geq 0$, and $k > 0$ is a constant. Here, $x'(0) = 0$ corresponds to the fact that the pellet is initially at rest for $t < 0$.

Taking the transform of both sides of the equation, we have

$$m\,s^2\,\mathcal{L}(x) + ks\,\mathcal{L}(x) = m\,v_0\,\mathcal{L}(\delta) = m\,v_0,$$

$$\mathcal{L}(x) = \frac{m\,v_0}{m\,s^2 + ks} = \frac{v_0}{s(s + k/m)}.$$

Writing

$$\frac{v_0}{s(s + k/m)} = \frac{A}{s} + \frac{B}{s + k/m},$$

we find that

$$A = \frac{m\,v_0}{k}, \qquad B = -\frac{m\,v_0}{k},$$

and

$$\mathcal{L}(x) = \frac{m\,v_0/k}{s} - \frac{m\,v_0/k}{s + k/m}.$$

The solution given by the inverse transform is

$$x(t) = \frac{m\,v_0}{k}\left(1 - e^{-\frac{k}{m}t}\right)$$

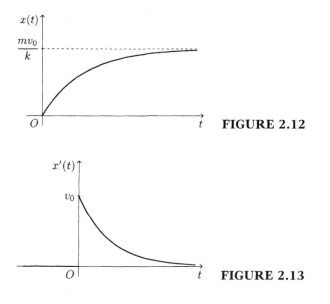

$x(t)$

$\frac{mv_0}{k}$

O t **FIGURE 2.12**

$x'(t)$

v_0

O t **FIGURE 2.13**

(Figure 2.12). Computing the velocity,

$$x'(t) = v_0\, e^{-\frac{k}{m}t},$$

and $\lim_{t\to 0^+} x'(t) = v_0$, whereas $\lim_{t\to 0^-} x'(t) = 0$, indicating the instantaneous jump in velocity at $t = 0$, from a rest state to the value v_0 (Figure 2.13).

Another formulation of this problem would be

$$m\frac{d^2x}{dt^2} + k\frac{dx}{dt} = 0, \qquad x(0) = 0, \ x'(0) = v_0.$$

Solving this version yields the same results as above.

Example 2.31. Suppose that at time $t = 0$ an impulse of $1V$ is applied to an RCL circuit (Figure 2.6), and for $t < 0$, $I(t) = 0$ and the charge on the capacitor is zero. This can be modeled by the equation

$$L\frac{dI}{dt} + RI + \frac{1}{C}\int_0^t I(\tau)\,d\tau = \delta(t),$$

where L, R, and C are positive constants, and

(i) $\dfrac{L}{C} > \dfrac{R^2}{4}$,   **(ii)** $\dfrac{L}{C} < \dfrac{R^2}{4}$.

Applying the Laplace transform gives

$$Ls\,\mathcal{L}(I) + R\,\mathcal{L}(I) + \frac{1}{Cs}\,\mathcal{L}(I) = 1,$$

that is,

$$\mathcal{L}(I) = \frac{s}{Ls^2 + Rs + 1/C}$$

$$= \frac{s}{L[(s + R/2L)^2 + (1/LC - R^2/4L^2)]}.$$

Setting $a = R/2L$, $b^2 = 1/LC - R^2/4L^2 > 0$, assuming (i), then,

$$L\,\mathcal{L}(I) = \frac{s}{(s + a)^2 + b^2}$$

$$= \frac{s + a}{(s + a)^2 + b^2} - \frac{a}{(s + a)^2 + b^2}, \qquad (2.42)$$

and so

$$I(t) = \frac{e^{-at}}{L}\left(\cos bt - \frac{a}{b}\sin bt\right).$$

Assuming (ii), (2.42) becomes

$$L\mathcal{L}(I) = \frac{s}{(s + a)^2 - b^2} = \frac{s + a}{(s + a)^2 - b^2} - \frac{a}{(s + a)^2 - b^2}$$

with $a = R/2L$, $b^2 = R^2/4L^2 - 1/LC > 0$. Consequently,

$$I(t) = \frac{e^{-at}}{L}\left(\cosh bt - \frac{a}{b}\sinh bt\right).$$

A Mechanical System. We consider a mass m suspended on a spring that is rigidly supported from one end (Figure 2.14). The rest position is denoted by $x = 0$, downward displacement is represented by $x > 0$, and upward displacement is shown by $x < 0$.
  To analyze this situation let

  i. $k > 0$ be the spring constant from Hooke's law,
  ii. $a(dx/dt)$ be the damping force due to the medium (e.g., air), where $a > 0$, that is, the damping force is proportional to the velocity,
  iii. $F(t)$ represents all external impressed forces on m.

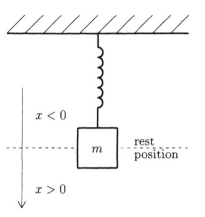

$x < 0$

m rest position

$x > 0$

FIGURE 2.14

Newton's second law states that the sum of the forces acting on m equals $m\, d^2x/dt^2$, that is,

$$m \frac{d^2x}{dt^2} = -kx - a \frac{dx}{dt} + F(t),$$

or

$$m \frac{d^2x}{dt^2} + a \frac{dx}{dt} + kx = F(t). \tag{2.43}$$

This equation is called the *equation of motion*.

Remark 2.32. If $a = 0$, the motion is called *undamped*. If $a \neq 0$, the motion is called *damped*. If $F(t) \equiv 0$ (i.e., no impressed forces), the motion is called *free*; otherwise it is *forced*.

We can write (2.43) with $F(t) \equiv 0$ as

$$\frac{d^2x}{dt^2} + \frac{a}{m} \frac{dx}{dt} + \frac{k}{m} x = 0.$$

Setting $a/m = 2b$, $k/m = \lambda^2$, we obtain

$$\frac{d^2x}{dt^2} + 2b \frac{dx}{dt} + \lambda^2 x = 0. \tag{2.44}$$

The characteristic equation is

$$r^2 + 2br + \lambda^2 = 0,$$

with roots

$$r = -b \pm \sqrt{b^2 - \lambda^2}.$$

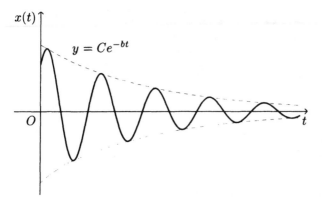

$x(t)$

$y = Ce^{-bt}$

O t

FIGURE 2.15

The resulting behavior of the system depends on the relation between b and λ. One interesting case is when $0 < b < \lambda$, where we obtain

$$x(t) = e^{-bt}(c_1 \sin\sqrt{\lambda^2 - b^2}\,t + c_2 \cos\sqrt{\lambda^2 - b^2}\,t),$$

which after some algebraic manipulations (setting $c = \sqrt{c_1^2 + c_2^2}$, $\cos\varphi = c_2/c$) becomes

$$x(t) = c\,e^{-bt}\cos(\sqrt{\lambda^2 - b^2}\,t - \varphi).$$

This represents the behavior of damped oscillation (Figure 2.15).

Let us apply a unit impulse force to the above situation.

Example 2.33.   For $0 < b < \lambda$, suppose that

$$\frac{d^2x}{dt^2} + 2b\frac{dx}{dt} + \lambda^2 x = \delta(t), \qquad x(0) = 0,\ x'(0) = 0,$$

which models the response of the mechanical system to a unit impulse.

Therefore,

$$\mathcal{L}(x'') + 2b\,\mathcal{L}(x') + \lambda^2\mathcal{L}(x) = \mathcal{L}(\delta) = 1,$$

so that

$$\mathcal{L}(x) = \frac{1}{s^2 + 2bs + \lambda^2}$$

$$= \frac{1}{(s+b)^2 + (\lambda^2 - b^2)},$$

and

$$x(t) = \frac{1}{\sqrt{\lambda^2 - b^2}}\, e^{-bt} \sin(\sqrt{\lambda^2 - b^2}\, t),$$

which again is a case of damped oscillation.

Exercises 2.5

1. Solve

$$\frac{d^2y}{dt^2} - 4\frac{dy}{dt} + 2y = \delta(t), \qquad y(0) = y'(0) = 0.$$

2. The response of a spring with no damping ($a = 0$) to a unit impulse at $t = 0$ is given by

$$m\frac{d^2x}{dt^2} + kx = \delta(t), \qquad x(0) = 0, \; x'(0) = 0.$$

Determine $x(t)$.

3. Suppose that the current in an RL circuit satisifies

$$L\frac{dI}{dt} + RI = E(t),$$

where L, and R are constants, and $E(t)$ is the impressed voltage. Find the response to a unit impulse at $t = 0$, assuming $E(t) = 0$ for $t \leq 0$.

4. Solve

$$m\frac{d^2x}{dt^2} + a\frac{dx}{dt} + kx = \delta(t),$$

for $m = 1$, $a = 2$, $k = 1$, $x(0) = x'(0) = 0$.

5. Show that if f satisfies the conditions of the derivative theorem (2.7), then

$$\mathcal{L}^{-1}\big(sF(s)\big) = f'(t) + f(0)\,\delta(t).$$

6. Show that

$$\mathcal{L}^{-1}\left(\frac{s-a}{s+a}\right) = \delta(t) - 2ae^{-at}.$$

7. A certain function $U(x)$ satisfies

$$a^2 U'' - b^2 U = -\frac{1}{2} \delta, \qquad x > 0,$$

where a and b are positive constants. If $U(x) \to 0$ as $x \to \infty$, and $U(-x) = U(x)$, show that

$$U(x) = \frac{1}{2ab} e^{-\frac{b}{a}|x|}.$$

[Hint: Take $U(0) = c$, $U'(0) = 0$, where c is to be determined.]

2.6 Asymptotic Values

Two properties of the Laplace transform are sometimes useful in determining limiting values of a function $f(t)$ as $t \to 0$ or as $t \to \infty$, even though the function is not known explicitly. This is achieved by examining the behavior of $\mathcal{L}(f(t))$.

Theorem 2.34 (Initial-Value Theorem). *Suppose that f, f' satisfy the conditions as in the derivative theorem* (2.7), *and* $F(s) = \mathcal{L}(f(t))$. *Then*

$$f(0^+) = \lim_{t \to 0^+} f(t) = \lim_{s \to \infty} s F(s) \qquad \text{(s real)}.$$

PROOF. By the general property of all Laplace transforms (of functions), we know that $\mathcal{L}(f'(t)) = G(s) \to 0$ as $s \to \infty$ (Theorem 1.20). By the derivative theorem,

$$G(s) = s F(s) - f(0^+), \qquad s > \alpha.$$

Taking the limit,

$$0 = \lim_{s \to \infty} G(s) = \lim_{s \to \infty} \left(s F(s) - f(0^+) \right).$$

Therefore,

$$f(0^+) = \lim_{s \to \infty} s F(s). \qquad \square$$

Example 2.35. If

$$\mathcal{L}(f(t)) = \frac{s+1}{(s-1)(s+2)},$$

then

$$f(0^+) = \lim_{s \to \infty} s \left(\frac{s+1}{(s-1)(s+2)} \right) = 1.$$

Theorem 2.36 (Terminal-Value Theorem). *Suppose that f satisfies the conditions of the derivative theorem (2.7) and furthermore that* $\lim_{t \to \infty} f(t)$ *exists. Then this limiting value is given by*

$$\lim_{t \to \infty} f(t) = \lim_{s \to 0} s F(s) \qquad (s \text{ real}),$$

where $F(s) = \mathcal{L}(f(t))$.

PROOF.   First note that f has exponential order $\alpha = 0$ since it is bounded in view of the hypothesis. By the derivative theorem,

$$G(s) = \mathcal{L}(f'(t)) = s F(s) - f(0^+) \qquad (s > 0).$$

Taking the limit,

$$\lim_{s \to 0} G(s) = \lim_{s \to 0} s F(s) - f(0^+). \tag{2.45}$$

Furthermore,

$$\lim_{s \to 0} G(s) = \lim_{s \to 0} \int_0^\infty e^{-st} f'(t) \, dt$$

$$= \int_0^\infty f'(t) \, dt, \tag{2.46}$$

since in this particular instance the limit can be passed inside the integral (see Corollary A.4). The integral in (2.46) exists since it is nothing but

$$\int_0^\infty f'(t) \, dt = \lim_{\tau \to \infty} \int_0^\tau f'(t) \, dt$$

$$= \lim_{\tau \to \infty} [f(\tau) - f(0^+)]. \tag{2.47}$$

Equating (2.45), (2.46), and (2.47),

$$\lim_{t \to \infty} f(t) = \lim_{s \to 0} s F(s). \qquad \qquad \square$$

Example 2.37.   Let $f(t) = \sin t$. Then

$$\lim_{s \to 0} s F(s) = \lim_{s \to 0} \frac{s}{s^2 + 1} = 0,$$

but $\lim_{t\to\infty} f(t)$ does not exist! Thus we may deduce that if $\lim_{s\to 0} s\,F(s) = L$, then either $\lim_{t\to\infty} f(t) = L$ or *this limit does not exist*. That is the best we can do without knowing a priori that $\lim_{t\to\infty} f(t)$ exists.

Exercises 2.6

1. Without determining $f(t)$ and assuming $f(t)$ satisfies the hypotheses of the initial-value theorem (2.34), compute $f(0^+)$ if

   (a) $\mathcal{L}(f(t)) = \dfrac{s^3 + 3a^2 s}{(s^2 - a^2)^3}$

   (b) $\mathcal{L}(f(t)) = \dfrac{\sqrt{s^2 + a^2} - s)^n}{\sqrt{s^2 + a^2}} \quad (n \geq 0)$

   (c) $\log\left(\dfrac{s+a}{s+b}\right) \quad (a \neq b).$

2. Without determining $f(t)$, and assuming $f(t)$ satisfies the hypotheses of the terminal-value theorem (2.36), compute $\lim_{t\to\infty} f(t)$ if

   (a) $\mathcal{L}(f(t)) = \dfrac{s+b}{(s+b)^2 + a^2} \quad (b > 0)$

   (b) $\mathcal{L}(f(t)) = \dfrac{1}{s} + \tan^{-1}\left(\dfrac{a}{s}\right).$

3. Show that

$$\lim_{s\to 0} s\,\frac{s}{(s^2 + a^2)^2}$$

   exists, and

$$\mathcal{L}\left(\frac{t}{2a}\sin at\right) = \frac{s}{(s^2 + a^2)^2},$$

   yet

$$\lim_{t\to\infty} \frac{t}{2a}\sin at$$

   does not exist.

2.7 Convolution

The convolution of two functions, $f(t)$ and $g(t)$, defined for $t > 0$, plays an important role in a number of different physical applications.

The convolution is given by the integral

$$(f * g)(t) = \int_0^t f(\tau)g(t - \tau)\,d\tau,$$

which of course exists if f and g are, say, piecewise continuous. Substituting $u = t - \tau$ gives

$$(f * g)(t) = \int_0^t g(u)f(t - u)\,du = (g * f)(t),$$

that is, the convolution is *commutative*.

Other basic properties of the convolution are as follows:

(i) $c(f * g) = cf * g = f * cg$, c constant;
(ii) $f * (g * h) = (f * g) * h$ (*associative property*);
(iii) $f * (g + h) = (f * g) + (f * h)$ (*distributive property*).

Properties (i) and (iii) are routine to verify. As for (ii),

$$[f * (g * h)](t)$$

$$= \int_0^t f(\tau)(g * h)(t - \tau)\,d\tau$$

$$= \int_0^t f(\tau)\left(\int_0^{t-\tau} g(x)h(t - \tau - x)\,dx\right)d\tau$$

$$= \int_0^t \left(\int_\tau^t f(\tau)g(u - \tau)h(t - u)\,du\right)d\tau \quad (x = u - \tau)$$

$$= \int_0^t \left(\int_0^u f(\tau)g(u - \tau)\,d\tau\right)h(t - u)\,du$$

$$= [(f * g) * h](t),$$

having reversed the order of integration.

Example 2.38. If $f(t) = e^t$, $g(t) = t$, then

$$(f * g)(t) = \int_0^t e^\tau (t - \tau) \, d\tau$$

$$= t e^\tau \Big|_0^t - (\tau e^\tau - e^\tau) \Big|_0^t$$

$$= e^t - t - 1.$$

One of the very significant properties possessed by the convolution in connection with the Laplace transform is that *the Laplace transform of the convolution of two functions is the product of their Laplace transforms.*

Theorem 2.39 (Convolution Theorem). *If f and g are piecewise continuous on $[0, \infty)$ and of exponential order α, then*

$$\mathcal{L}[(f * g)(t)] = \mathcal{L}(f(t)) \cdot \mathcal{L}(g(t)) \qquad (\mathcal{R}e(s) > \alpha).$$

PROOF.   Let us start with the product

$$\mathcal{L}(f(t)) \cdot \mathcal{L}(g(t)) = \left(\int_0^\infty e^{-s\tau} f(\tau) \, d\tau \right) \left(\int_0^\infty e^{-su} g(u) \, du \right)$$

$$= \int_0^\infty \left(\int_0^\infty e^{-s(\tau + u)} f(\tau) g(u) \, du \right) d\tau.$$

Substituting $t = \tau + u$, and noting that τ is fixed in the interior integral, so that $du = dt$, we have

$$\mathcal{L}(f(t)) \cdot \mathcal{L}(g(t)) = \int_0^\infty \left(\int_\tau^\infty e^{-st} f(\tau) g(t - \tau) \, dt \right) d\tau. \qquad (2.48)$$

If we define $g(t) = 0$ for $t < 0$, then $g(t - \tau) = 0$ for $t < \tau$ and we can write (2.48) as

$$\mathcal{L}(f(t)) \cdot \mathcal{L}(g(t)) = \int_0^\infty \int_0^\infty e^{-st} f(\tau) g(t - \tau) \, dt \, d\tau.$$

Due to the hypotheses on f and g, the Laplace integrals of f and g converge absolutely and hence, in view of the preceding calculation,

$$\int_0^\infty \int_0^\infty \left| e^{-st} f(\tau) g(t - \tau) \right| \, dt \, d\tau$$

converges. This fact allows us to reverse the order of integration,* so that

$$\mathcal{L}\big(f(t)\big) \cdot \mathcal{L}\big(g(t)\big) = \int_0^\infty \int_0^\infty e^{-st} f(\tau) g(t - \tau) \, d\tau \, dt$$

$$= \int_0^\infty \left(\int_0^t e^{-st} f(\tau) g(t - \tau) \, d\tau \right) dt$$

$$= \int_0^\infty e^{-st} \left(\int_0^t f(\tau) g(t - \tau) \, d\tau \right) dt$$

$$= \mathcal{L}[(f * g)(t)]. \qquad \qquad \square$$

Example 2.40.

$$\mathcal{L}(e^{at} * e^{bt}) = \frac{1}{(s - a)(s - b)}.$$

Moreover,

$$\mathcal{L}^{-1} \left(\frac{1}{(s - a)(s - b)} \right) = e^{at} * e^{bt}$$

$$= \int_0^t e^{a\tau} e^{b(t-\tau)} d\tau$$

$$= \frac{e^{at} - e^{bt}}{a - b} \qquad a \neq b.$$

---

*Let

$$a_{mn} = \int_n^{n+1} \int_m^{m+1} |h(t, \tau)| \, dt \, d\tau, \qquad b_{mn} = \int_n^{n+1} \int_m^{m+1} h(t, \tau) \, dt \, d\tau,$$

so that $|b_{mn}| \leq a_{mn}$. If

$$\int_0^\infty \int_0^\infty |h(t, \tau)| \, dt \, d\tau < \infty,$$

then $\sum_{n=0}^\infty \sum_{m=0}^\infty a_{mn} < \infty$, implying $\sum_{n=0}^\infty \sum_{m=0}^\infty |b_{mn}| < \infty$. Hence, by a standard result on double series, the order of summation can be interchanged:

$$\sum_{n=0}^\infty \sum_{m=0}^\infty b_{mn} = \sum_{m=0}^\infty \sum_{n=0}^\infty b_{mn},$$

i.e.,

$$\int_0^\infty \int_0^\infty h(t, \tau) \, dt \, d\tau = \int_0^\infty \int_0^\infty h(t, \tau) \, d\tau \, dt.$$

Example 2.41. Find

$$\mathcal{L}^{-1}\left(\frac{1}{s^2(s-1)}\right).$$

Previously we applied a partial fraction decomposition. But we can also write

$$\frac{1}{s^2(s-1)} = \frac{1}{s^2} \cdot \frac{1}{s-1},$$

where $\mathcal{L}(t) = 1/s^2$, $\mathcal{L}(e^t) = 1/(s-1)$. By the convolution theorem,

$$\frac{1}{s^2} \cdot \frac{1}{s-1} = \mathcal{L}(t * e^t),$$

and thus

$$\mathcal{L}^{-1}\left(\frac{1}{s^2(s-1)}\right) = t * e^t$$
$$= e^t - t - 1$$

by Example 2.38.

Example 2.42.

(i) $\dfrac{\omega^2}{(s^2+\omega^2)^2} = \dfrac{\omega}{s^2+\omega^2} \cdot \dfrac{\omega}{s^2+\omega^2}$

$$= \mathcal{L}(\sin\omega t * \sin\omega t),$$

so that

$$\mathcal{L}^{-1}\left(\frac{\omega^2}{(s^2+\omega^2)^2}\right) = \sin\omega t * \sin\omega t$$
$$= \int_0^t \sin w\tau \sin\omega(t-\tau)\,d\tau$$
$$= \frac{1}{2w}(\sin\omega t - \omega t \cos\omega t).$$

Similarly,

(ii) $\mathcal{L}^{-1}\left(\dfrac{s}{(s^2+\omega^2)^2}\right) = \dfrac{1}{\omega}\cos\omega t * \sin\omega t$

$$= \frac{1}{\omega}\int_0^t \cos\omega\tau \sin\omega(t-\tau)\,d\tau$$

$$= \frac{1}{2\omega}t\sin\omega t.$$

Here we have used the fact that

$$\sin(A-B) = \sin A\cos B - \cos A\sin B$$

to compute both integrals.

These examples illustrate the utility of the convolution theorem in evaluating inverse transforms that are products.

Error Function. The error function from the theory of probability is defined as

$$\operatorname{erf}(t) = \frac{2}{\sqrt{\pi}}\int_0^t e^{-x^2}\,dx.$$

Note that

$$\lim_{t\to\infty}\operatorname{erf}(t) = \frac{2}{\sqrt{\pi}}\int_0^\infty e^{-x^2}\,dx = 1 \tag{2.49}$$

by Example 2.1 (Figure 2.16). The error function is related to Laplace transforms through the problem (see also Chapters 4 and 5) of finding

$$\mathcal{L}^{-1}\left(\frac{1}{\sqrt{s}(s-1)}\right).$$

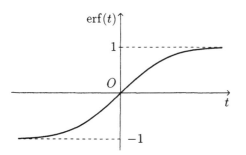

FIGURE 2.16

We know from (2.5) that

$$\mathcal{L}\left(\frac{1}{\sqrt{\pi t}}\right) = \frac{1}{\sqrt{s}}$$

and also that $\mathcal{L}(e^t) = 1/(s - 1)$. Then by the convolution theorem,

$$\mathcal{L}^{-1}\left(\frac{1}{\sqrt{s}(s-1)}\right) = \frac{1}{\sqrt{\pi t}} * e^t$$

$$= \int_0^t \frac{1}{\sqrt{\pi x}} e^{t-x} dx$$

$$= \frac{e^t}{\sqrt{\pi}} \int_0^t \frac{e^{-x}}{\sqrt{x}} dx.$$

Substituting $u = \sqrt{x}$ gives

$$\mathcal{L}^{-1}\left(\frac{1}{\sqrt{s}(s-1)}\right) = \frac{2e^t}{\sqrt{\pi}} \int_0^{\sqrt{t}} e^{-u^2} du$$

$$= e^t \operatorname{erf}(\sqrt{t}).$$

Applying the first translation theorem 1.27 with $a = -1$ yields

$$\mathcal{L}\left(\operatorname{erf}(\sqrt{t})\right) = \frac{1}{s\sqrt{s+1}}.$$

Beta Function. If $f(t) = t^{a-1}$, $g(t) = t^{b-1}$, $a, b > 0$, then

$$(f * g)(t) = \int_0^t \tau^{a-1}(t - \tau)^{b-1} d\tau.$$

Substituting $\tau = ut$,

$$(f * g)(t) = t^{a+b-1} \int_0^1 u^{a-1}(1 - u)^{b-1} du. \qquad (2.50)$$

The term

$$B(a, b) = \int_0^1 u^{a-1}(1 - u)^{b-1} du \qquad (2.51)$$

is known as the *beta function*. Then by the convolution theorem,

$$\mathcal{L}\left(t^{a+b-1} B(a, b)\right) = \mathcal{L}(t^{a-1})\, \mathcal{L}(t^{b-1})$$

$$= \frac{\Gamma(a)\,\Gamma(b)}{s^{a+b}}$$

by (2.2). Therefore,

$$t^{a+b-1}B(a, b) = \mathcal{L}^{-1}\left(\frac{\Gamma(a)\Gamma(b)}{s^{a+b}}\right)$$

$$= \Gamma(a)\Gamma(b)\frac{t^{a+b-1}}{\Gamma(a+b)}, \tag{2.52}$$

and we obtain Euler's formula for the beta function:

$$B(a, b) = \frac{\Gamma(a)\Gamma(b)}{\Gamma(a+b)}. \tag{2.53}$$

Calculating $B(1/2, 1/2)$ in (2.51) with $u = \sin^2\theta$, we find from (2.53)

$$\pi = B\left(\tfrac{1}{2}, \tfrac{1}{2}\right) = \left[\Gamma\left(\tfrac{1}{2}\right)\right]^2,$$

that is,

$$\Gamma\left(\tfrac{1}{2}\right) = \sqrt{\pi} \tag{2.54}$$

since $\Gamma(1/2) > 0$. See also Example 2.1.

Bessel Function. This important function is the solution to the Bessel equation of order v,

$$t^2\frac{d^2y}{dt^2} + t\frac{dy}{dt} + (t^2 - v^2)y = 0, \tag{2.55}$$

and is given by $\big($the solution to (2.55) has $a = 1\big)$

$$J_v(at) = \sum_{n=0}^{\infty}\frac{(-1)^n(at)^{2n+v}}{2^{2n+v}n!(n+v)!},$$

where $(n+v)! = \Gamma(n+v+1)$. For $v = 0$,

$$J_0(at) = \sum_{n=0}^{\infty}\frac{(-1)^n a^{2n}t^{2n}}{2^{2n}(n!)^2} = \sum_{n=0}^{\infty}a_{2n}t^{2n}$$

(Figure 2.17). $J_0(at)$ is a bounded function and

$$|a_{2n}| = \frac{|a|^{2n}}{2^{2n}(n!)^2} \le \frac{|a|^{2n}}{(2n)!}.$$

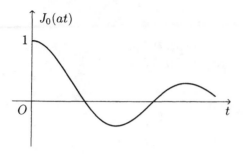

$J_0(at)$

FIGURE 2.17

The latter inequality for $n = 0, 1, 2, \cdots$ can be verified by induction. Taking $\alpha = |a|$ in Theorem 1.18 means that we can take the Laplace transform of $J_0(at)$ term by term.

Hence,

$$\mathcal{L}\big(J_0(at)\big) = \sum_{n=0}^{\infty} \frac{(-1)^n a^{2n}}{2^{2n}(n!)^2} \mathcal{L}(t^{2n})$$

$$= \sum_{n=0}^{\infty} \frac{(-1)^n a^{2n}(2n)!}{2^{2n}(n!)^2 s^{2n+1}}$$

$$= \frac{1}{s} \sum_{n=0}^{\infty} \frac{(-1)^n (2n)!}{2^{2n}(n!)^2} \left(\frac{a^2}{s^2}\right)^n$$

$$= \frac{1}{s} \left(\frac{s}{\sqrt{s^2 + a^2}}\right) \qquad (\mathcal{R}e(s) > |a|)$$

$$= \frac{1}{\sqrt{s^2 + a^2}}.$$

Here we have used the Taylor series expansion

$$\frac{1}{\sqrt{1 + x^2}} = \sum_{n=0}^{\infty} \frac{(-1)^n (2n)!}{2^{2n}(n!)^2} x^{2n} \qquad (|x| < 1)$$

with $x = a/s$.

Integral Equations. Equations of the form

$$f(t) = g(t) + \int_0^t k(t, \tau) f(\tau) \, d\tau$$

and

$$g(t) = \int_0^t k(t, \tau) f(\tau) d\tau$$

are known as *integral equations*, where $f(t)$ is the unknown function. When the *kernel* $k(t, \tau)$ is of the particular form

$$k(t, \tau) = k(t - \tau),$$

the integrals represent convolutions. In this case, the Laplace transform lends itself to their solution.

Considering the first type, if g and k are known, then formally

$$\mathcal{L}(f) = \mathcal{L}(g) + \mathcal{L}(f) \mathcal{L}(k)$$

by the convolution theorem. Then

$$\mathcal{L}(f) = \frac{\mathcal{L}(g)}{1 - \mathcal{L}(k)},$$

and from this expression $f(t)$ often can be found since the right-hand side is just a function of the variable s.

Example 2.43. Solve the integral equation

$$x(t) = e^{-t} + \int_0^t \sin(t - \tau) x(\tau) d\tau.$$

We apply the Laplace transform to both sides of the equation so that

$$\mathcal{L}(x(t)) = \mathcal{L}(e^{-t}) + \mathcal{L}(\sin t) \mathcal{L}(x(t))$$

and

$$\mathcal{L}(x(t)) = \frac{\mathcal{L}(e^{-t})}{1 - \mathcal{L}(\sin t)}$$

$$= \frac{s^2 + 1}{s^2(s + 1)}$$

$$= \frac{2}{s + 1} + \frac{1}{s^2} - \frac{1}{s}.$$

Hence

$$x(t) = 2e^{-t} + t - 1.$$

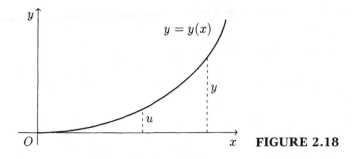

$y = y(x)$

O $\qquad$ x $\qquad$ **FIGURE 2.18**

As an example of an integral equation of the second type, let us consider a classical problem from the 19th century. A particle is to slide down a frictionless curve with the requirement that the duration of descent (due to gravity) is independent of the starting point (Figure 2.18). Such a curve is called a *tautochrone*.

An analysis of the physics of the situation leads to the (Abel) integral equation

$$T_0 = \frac{1}{\sqrt{2g}} \int_0^y \frac{f(u)\,du}{\sqrt{y-u}}, \qquad (2.56)$$

where T_0 is a constant (time), g is the gravitational constant, and $f(u)$ represents ds/dy at $y = u$, where $s = s(y)$ is arc length. The integral (2.56) then is the convolution of the functions $f(y)$ and $1/\sqrt{y}$.

Taking the transform gives

$$\mathcal{L}(T_0) = \frac{1}{\sqrt{2g}} \mathcal{L}(f(y)) \mathcal{L}\left(\frac{1}{\sqrt{y}}\right),$$

and so by (2.4)

$$\mathcal{L}(f(y)) = \frac{\sqrt{2g}\,T_0/s}{\sqrt{\pi/s}}.$$

Therefore,

$$\mathcal{L}(f(y)) = \frac{\sqrt{2g/\pi}}{s^{\frac{1}{2}}}\,T_0 = \frac{c_0}{s^{\frac{1}{2}}}.$$

The inverse transform gives

$$f(y) = \frac{c}{\sqrt{y}}. \qquad (2.57)$$

Since

$$f(y) = \frac{ds}{dy} = \sqrt{1 + \left(\frac{dx}{dy}\right)^2},$$

we arrive at the differential equation

$$1 + \left(\frac{dx}{dy}\right)^2 = \frac{c^2}{y},$$

invoking (2.57). Then

$$x = \int \sqrt{\frac{c^2 - y}{y}}\, dy.$$

Setting $y = c^2 \sin^2(\varphi/2)$ leads to

$$x = \frac{c^2}{2}(\varphi + \sin\varphi), \qquad y = \frac{c^2}{2}(1 - \cos\varphi),$$

which are the parametric equations of a *cycloid*.

Exercises 2.7

1. Use the convolution theorem to find the following:

   (a) $\mathcal{L}^{-1}\left(\dfrac{1}{(s-1)(s+2)}\right)$

   (b) $\mathcal{L}^{-1}\left(\dfrac{1}{s(s^2+1)}\right)$

   (c) $\mathcal{L}^{-1}\left(\dfrac{1}{s^2(s^2+1)}\right)$

   (d) $\mathcal{L}^{-1}\left(\dfrac{1}{s^2(s+4)^2}\right)$

   (e) $\mathcal{L}^{-1}\left(\dfrac{s}{(s^2+1)^3}\right).$

2. Prove the distributive property for convolutions:

   $$f * (g + h) = f * g + f * h.$$

3. Show that if f and g are piecewise continuous and of exponential order on $[0, \infty)$, then $(f * g)(t)$ is of exponential order on $[0, \infty)$.

4. Use the convolution theorem to show that

$$\int_0^t \cos\tau \, \sin(t-\tau) \, d\tau = \tfrac{1}{2} t \sin t.$$

5. Show that

$$y(t) = \frac{1}{\omega} \int_0^t f(\tau) \sinh\omega(t-\tau) \, d\tau$$

is a solution to the differential equation

$$y'' - \omega^2 y = f(t), \qquad y(0) = y'(0) = 0,$$

for f continuous of exponential order on $[0,\infty)$.

6. Determine

(a) $\mathcal{L}^{-1}\left(\dfrac{1}{s\sqrt{s+a}}\right)$ (b) $\mathcal{L}(e^{at}\,\mathrm{erf}\sqrt{at})$

(c) $\mathcal{L}(t\,\mathrm{erf}\sqrt{at})$.

7. Find

(a) $\mathcal{L}^{-1}\left(\dfrac{e^{-s}}{\sqrt{s^2+1}}\right)$ (b) $\mathcal{L}^{-1}\left(\dfrac{1}{s\sqrt{s^2+a^2}}\right)$.

8. Evaluate $\displaystyle\int_0^1 u^{-\frac{1}{2}}(1-u)^{\frac{1}{2}}\,du$.

9. The *modified Bessel function of order* v is given by $I_v(t) = i^{-v}J_v(it) = \sum_{n=0}^{\infty} t^{2n+v}/2^{2n+v}n!(n+v)!$. Show that

$$\mathcal{L}\big(I_0(at)\big) = \frac{1}{\sqrt{s^2-a^2}} \qquad (\mathcal{R}e(s) > |a|).$$

10. Solve the following integral equations:

(a) $x(t) = 1 + \displaystyle\int_0^t \cos(t-\tau)x(\tau)\,d\tau$

(b) $x(t) = \sin t + \displaystyle\int_0^t e^\tau x(t-\tau)\,d\tau$

(c) $x(t) = \displaystyle\int_0^t (\sin\tau)x(t-\tau)\,d\tau$

(d) $te^{-at} = \displaystyle\int_0^t x(\tau)x(t-\tau)\,d\tau$.

11. Solve the integro-differential equations

(a) $x'(t) + \int_0^t x(t - \tau)\, d\tau = \cos t,$ $\qquad x(0) = 0$

(b) $\sin t = \int_0^t x''(\tau)\, x(t - \tau)\, d\tau,$ $\qquad x(0) = x'(0) = 0.$

12. Solve the initial-value problem

$$y'' - 2y' - 3y = f(t), \qquad y(0) = y'(0) = 0.$$

2.8 Steady-State Solutions

Let us consider the general nth-order, linear, nonhomogeneous differential equation with constant coefficients

$$y^{(n)} + a_{n-1}y^{(n-1)} + \cdots + a_1 y' + a_0 y = f(t) \qquad (2.58)$$

for $f \in L$, and with initial conditions

$$y(0) = y'(0) = \cdots = y^{(n-1)}(0) = 0. \qquad (2.59)$$

To be more precise, we should really say

$$y(0^+) = y'(0^+) = \cdots = y^{(n-1)}(0^+) = 0,$$

but we shall continue to employ the conventional notation of (2.59).

A solution of (2.58) satisfying (2.59) is called a *steady-state solution*. By (2.22), proceeding formally,

$$\mathcal{L}\big(y^{(k)}(t)\big) = s^k \mathcal{L}\big(y(t)\big), \qquad k = 0, 1, 2, \ldots.$$

Thus, the Laplace transform of (2.58) is

$$(s^n + a_{n-1}s^{n-1} + \cdots + a_1 s + a_0)\mathcal{L}\big(y(t)\big) = \mathcal{L}\big(f(t)\big),$$

or

$$\mathcal{L}\big(y(t)\big) = \frac{\mathcal{L}\big(f(t)\big)}{Q(s)}, \qquad (2.60)$$

where $Q(s) = s^n + a_{n-1}s^{n-1} + \cdots + a_1 s + a_0$.

Suppose that

$$\frac{1}{Q(s)} = \mathcal{L}\big(g(t)\big)$$

for some function $g(t)$. Then

$$\mathcal{L}(y(t)) = \mathcal{L}(f(t))\,\mathcal{L}(g(t))$$
$$= \mathcal{L}[(f * g)(t)]$$

and

$$y(t) = \int_0^t f(\tau)g(t - \tau)\,d\tau = \int_0^t g(\tau)f(t - \tau)\,d\tau. \qquad (2.61)$$

Since

$$Q(s)\,\mathcal{L}(g(t)) = 1,$$

in other words,

$$(s^n + a_{n-1}s^{n-1} + \cdots + a_1 s + a_0)\,\mathcal{L}(g(t)) = \mathcal{L}(\delta(t)),$$

we may consider $g = g(t)$ to be the steady-state solution of

$$g^{(n)} + a_{n-1}g^{(n-1)} + \cdots + a_1 g' + a_0 g = \delta(t). \qquad (2.62)$$

This means that we can determine the solution $y = y(t)$ via (2.61) by first determining $g = g(t)$ as a steady-state solution of (2.62). In this case, $g(t)$ is known as the *impulsive response* since we are determining the response of the system (2.58) for $f(t) = \delta(t)$.

Example 2.44. Find the steady-state solution to

$$y'' - y = f(t) = e^{2t}$$

by first determining the response of the system to the Dirac delta function.

For $g'' - g = \delta(t)$,

$$s^2\mathcal{L}(g) - \mathcal{L}(g) = 1,$$

namely,

$$\mathcal{L}(g) = \frac{1}{s^2 - 1} = \frac{1/2}{s - 1} - \frac{1/2}{s + 1},$$

so that

$$g(t) = \frac{1}{2}e^t - \frac{1}{2}e^{-t}.$$

By (2.61),

$$y(t) = \int_0^t \left(\frac{1}{2} e^\tau - \frac{1}{2} e^{-\tau} \right) e^{2(t-\tau)} d\tau$$

$$= \frac{1}{2} e^{2t} \int_0^t (e^{-\tau} - e^{-3\tau}) d\tau$$

$$= \frac{1}{3} e^{2t} - \frac{1}{2} e^t + \frac{1}{6} e^{-t}.$$

This approach, while seemingly cumbersome, comes into its own when the impulse response $g(t)$ is not known explicitly, but only indirectly by experimental means.

By the same token, it is also worthwhile to determine the response of the steady-state system (2.58)/(2.59) to the unit step function, $u(t)$.

To this end, if $f(t) = u(t)$, the (indicial) response $h = h(t)$ satisfies

$$h^{(n)} + a_{n-1} h^{(n-1)} + \cdots + a_1 h' + a_0 h = u(t),$$
$$h(0) = h'(0) = \cdots = h^{(n-1)}(0) = 0.$$

(2.63)

Moreover,

$$\mathcal{L}(h(t)) = \frac{\mathcal{L}(u(t))}{Q(s)} = \frac{1}{s\, Q(s)}.$$

Revisiting (2.60) with $1/Q(s) = s\, \mathcal{L}(h(t))$,

$$\mathcal{L}(y(t)) = s\, \mathcal{L}(h(t))\, \mathcal{L}(f(t))$$
$$= \mathcal{L}(h'(t))\, \mathcal{L}(f(t)) \qquad (h(0) = 0)$$
$$= \mathcal{L}[(h' * f)(t)].$$

Therefore,

$$y(t) = \int_0^t h'(\tau) f(t-\tau)\, d\tau = \int_0^t f(\tau) h'(t-\tau)\, d\tau.$$

(2.64)

Once again we find that the steady-state solution of (2.58)/(2.59) can be determined by a convolution of a particular function with the input, $f(t)$, in this case the steady-state solution $h(t)$ of (2.63).

Note that in the preceding we could have witten

$$\mathcal{L}(y(t)) = s\, \mathcal{L}(f(t))\, \mathcal{L}(h(t))$$

$$= [\mathcal{L}(f'(t)) + f(0)] \mathcal{L}(h(t)),$$

and consequently

$$y(t) = \int_0^t f'(\tau) h(t - \tau) \, d\tau + f(0) h(t). \tag{2.65}$$

This approach involving the convolution with the indicial response is known as the *superposition principle*.

Example 2.45.　For the steady-state problem in Example 2.44 and $f(t) = u(t)$,

$$s^2 \mathcal{L}(h(t)) - \mathcal{L}(h(t)) = \mathcal{L}(u(t)) = \frac{1}{s},$$

that is,

$$\mathcal{L}(h(t)) = \frac{1}{s(s-1)(s+1)} = -\frac{1}{s} + \frac{1/2}{s-1} + \frac{1/2}{s+1},$$

and

$$h(t) = -1 + \frac{1}{2} e^t + \frac{1}{2} e^{-t},$$

$$h'(t) = \frac{1}{2} e^t - \frac{1}{2} e^{-t},$$

the latter quantity being exactly the expression obtained for $g(t)$ in the previous example. Then

$$y(t) = \int_0^t h'(\tau) f(t - \tau) \, d\tau$$

$$= \frac{1}{3} e^{2t} - \frac{1}{2} e^t + \frac{1}{6} e^{-t},$$

as before.

Let us go back to the polynomial $Q(s)$ of (2.60) and suppose that all of its roots $\alpha_1, \alpha_2, \cdots, \alpha_n$ are simple, so that we have the partial fraction decomposition

$$\frac{1}{Q(s)} = \sum_{k=1}^n \frac{A_k}{s - \alpha_k} = \mathcal{L}\left(\sum_{k=1}^n A_k e^{\alpha_k t} \right).$$

Putting this expression into (2.60) gives

$$\mathcal{L}\big(y(t)\big) = \mathcal{L}\big(f(t)\big) \mathcal{L}\left(\sum_{k=1}^{n} A_k e^{\alpha_k t}\right),$$

and so

$$y(t) = \int_0^t f(\tau) \sum_{k=1}^{n} A_k e^{\alpha_k(t-\tau)} d\tau$$

$$= \sum_{k=1}^{n} A_k \int_0^t f(\tau) e^{\alpha_k(t-\tau)} d\tau. \tag{2.66}$$

Since

$$A_k = \lim_{s \to \alpha_k} \frac{s - \alpha_k}{Q(s)},$$

we can write

$$A_k = \lim_{s \to \alpha_k} \frac{s - \alpha_k}{Q(s) - Q(\alpha_k)}$$

$$= \frac{1}{Q'(\alpha_k)}, \qquad k = 1, 2, \ldots, n,$$

invoking l'Hôpital's rule [$Q'(\alpha_k) \neq 0$ from Section 3.4 since the α_ks are simple].

A fortiori (2.66) becomes

$$y(t) = \sum_{k=1}^{n} \frac{1}{Q'(\alpha_k)} \int_0^t f(\tau) e^{\alpha_k(t-\tau)} d\tau, \tag{2.67}$$

the *Heaviside expansion theorem*.

Exercises 2.8

1. Solve the following steady-state problems by first determining the response of the system to the Dirac delta function and then using equation (2.61):

   (a) $y'' + y' - 2y = 4e^{-t}$

(b) $y'' - y = \sin t$
(c) $y''' - 2y'' - 5y' + 6y = 2t$.

2. Solve parts (a), (b), and (c) of Question 1 by first determining the response of the system to the unit step function and then using equation (2.64) or (2.65).

3. Solve parts (a), (b), and (c) of Question 1 directly by using the Heaviside expansion theorem (2.67).

4. A mass attached to a vertical spring undergoes forced vibration with damping so that the motion is given by

$$\frac{d^2 x}{dt^2} + \frac{dx}{dt} + 2x = \sin t,$$

where $x(t)$ is the displacement at time t. Determine the displacement at time t of the steady-state solution.

2.9 Difference Equations

A *difference equation* expresses the relationship between the values of a function $y(t)$ and the values of the function at different arguments, $y(t + h)$, h constant. For example,

$$y(t - 1) - 3y(t) + 2y(t - 2) = e^t,$$

$$y(t + 1)y(t) = \cos t$$

are difference equations, *linear* and *nonlinear*, respectively.

Equations that express the relationship between the terms of a sequence $a_0, a_1, a_2, \ldots$ are also difference equations, as, for example,

$$a_{n+2} - 3a_{n+1} + 2a_n = 5^n \quad \text{(linear)},$$

$$a_{n+1} = 2a_n^2 \quad \text{(nonlinear)}.$$

As we will see, the function and sequence forms are not as unrelated as they may appear, with the latter easily expressed in terms of the former. Both of the above linear forms are amenable to solution by the Laplace transform method.

For further reading on difference equations, see Mickens [8].

Example 2.46. Solve

$$y(t) - y\left(t - \frac{\pi}{\omega}\right) = \sin \omega t, \qquad y(t) = 0, \ t \le 0.$$

We compute

$$\mathcal{L}\left(y\left(t - \frac{\pi}{\omega}\right)\right) = \int_0^\infty e^{-st} y\left(t - \frac{\pi}{\omega}\right) dt$$

$$= \int_{-\frac{\pi}{\omega}}^\infty e^{-s\left(\tau + \frac{\pi}{\omega}\right)} y(\tau) \, d\tau \qquad \left(\tau = t - \frac{\pi}{\omega}\right)$$

$$= e^{-\frac{\pi s}{\omega}} \int_0^\infty e^{-s\tau} y(\tau) \, d\tau$$

$$= e^{-\frac{\pi s}{\omega}} \mathcal{L}(y(t)).$$

Therefore, taking the Laplace transform of both sides of the difference equation,

$$\mathcal{L}(y(t)) - e^{-\frac{\pi s}{\omega}} \mathcal{L}(y(t)) = \frac{\omega}{s^2 + \omega^2},$$

or

$$\mathcal{L}(y(t)) = \frac{\omega}{(s^2 + \omega^2)\left(1 - e^{-\frac{\pi s}{\omega}}\right)},$$

and

$$y(t) = \begin{cases} \sin \omega t & \frac{2n\pi}{\omega} < t < \frac{(2n+1)\pi}{\omega} \\ 0 & \frac{(2n+1)\pi}{\omega} < t < \frac{(2n+2)\pi}{\omega} \end{cases} \qquad n = 0, 1, 2, \ldots,$$

the half–wave-rectified sine function given in Example 2.6.

In order to solve difference equations that are in sequence form, the following result proves instrumental.

Example 2.47. $f(t) = a^{[t]}$, where $[t]$ is the greatest integer $\le t$, $a > 0$ (Figure 2.19). Then $f(t)$ has exponential order (Exercises 2.9, Question 1) and

$$\mathcal{L}(f(t)) = \int_0^\infty e^{-st} f(t) \, dt$$

$$= \int_0^1 e^{-st} a^0 dt + \int_1^2 e^{-st} a^1 dt + \int_2^3 e^{-st} a^2 dt + \cdots$$

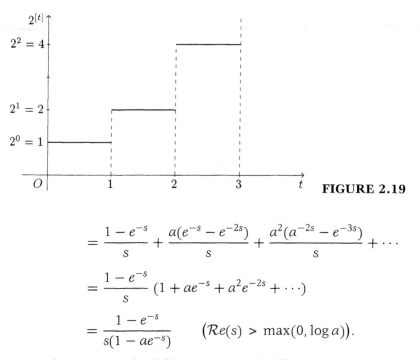

FIGURE 2.19

$$= \frac{1-e^{-s}}{s} + \frac{a(e^{-s} - e^{-2s})}{s} + \frac{a^2(a^{-2s} - e^{-3s})}{s} + \cdots$$

$$= \frac{1-e^{-s}}{s}(1 + ae^{-s} + a^2e^{-2s} + \cdots)$$

$$= \frac{1-e^{-s}}{s(1-ae^{-s})} \qquad (\mathcal{R}e(s) > \max(0, \log a)).$$

Let us then turn to the following type of difference equation.

Example 2.48. Solve

$$a_{n+2} - 3a_{n+1} + 2a_n = 0, \qquad a_0 = 0, \ a_1 = 1.$$

To treat this sort of problem, let us define

$$y(t) = a_n, \qquad n \le t < n+1, \ n = 0, 1, 2, \ldots.$$

Then our difference equation becomes

$$y(t+2) - 3y(t+1) + 2y(t) = 0. \qquad (2.68)$$

Taking the Laplace transform, we first have

$$\mathcal{L}(y(t+2)) = \int_0^\infty e^{-st}y(t+2)\,dt$$

$$= \int_2^\infty e^{-s(\tau-2)}y(\tau)\,d\tau \qquad (\tau = t+2)$$

$$= e^{2s}\int_0^\infty e^{-s\tau}y(\tau)\,d\tau - e^{2s}\int_0^2 e^{-s\tau}y(\tau)\,d\tau$$

$$= e^{2s} \mathcal{L}(y(t)) - e^{2s} \int_0^1 e^{-s\tau} a_0 d\tau - e^{2s} \int_1^2 e^{-s\tau} a_1 d\tau$$

$$= e^{2s} \mathcal{L}(y(t)) - e^{2s} \left(\frac{e^{-s} - e^{-2s}}{s} \right)$$

$$= e^{2s} \mathcal{L}(y(t)) - \frac{e^s}{s}(1 - e^{-s}),$$

since $a_0 = 0$, $a_1 = 1$.

Similarly,

$$\mathcal{L}(y(t+1)) = e^s \mathcal{L}(y(t)).$$

Thus the transform of (2.68) becomes

$$e^{2s} \mathcal{L}(y(t)) - \frac{e^s}{s}(1 - e^{-s}) - 3e^s \mathcal{L}(y(t)) + 2\mathcal{L}(y(t)) = 0,$$

or

$$\mathcal{L}(y(t)) = \frac{e^s(1 - e^{-s})}{s(e^{2s} - 3e^s + 2)}$$

$$= \frac{e^s(1 - e^{-s})}{s} \left(\frac{1}{e^s - 2} - \frac{1}{e^s - 1} \right)$$

$$= \frac{1 - e^{-s}}{s} \left(\frac{1}{1 - 2e^{-s}} - \frac{1}{1 - e^{-s}} \right)$$

$$= \frac{1 - e^{-s}}{s(1 - 2e^{-s})} - \frac{1 - e^{-s}}{s(1 - e^{-s})}$$

$$= \mathcal{L}(2^{[t]}) - \mathcal{L}(1)$$

by Example 2.47. The solution is then given by (equating the expressions for $y(t)$)

$$a_n = 2^n - 1, \qquad n = 0, 1, 2, \ldots.$$

Checking this result: $a_{n+2} = 2^{n+2} - 1$, $a_{n+1} = 2^{n+1} - 1$, and so

$$(2^{n+2} - 1) - 3(2^{n+1} - 1) + 2(2^n - 1) = 2^{n+2} - 3 \cdot 2^{n+1} + 2 \cdot 2^n$$

$$= 2 \cdot 2^{n+1} - 3 \cdot 2^{n+1} + 2^{n+1}$$

$$= 0,$$

as desired.

If in the preceding example the right-hand side had been something other than 0, say

$$a_{n+2} - 3a_{n+1} + 2a_n = 3^n, \qquad a_0 = 0, \ a_1 = 1,$$

it would have transpired that

$$\mathcal{L}(y(t)) = \mathcal{L}(2^{[t]}) - \mathcal{L}(1) + \frac{\mathcal{L}(3^{[t]})}{e^{2s} - 3e^s + 2}$$

and

$$\frac{\mathcal{L}(3^{[t]})}{e^{2s} - 3e^s + 2} = \frac{1 - e^{-s}}{s(1 - 3e^{-s})} \frac{1}{e^{3s} - 3e^s + 2}$$

$$= \frac{e^s - 1}{s(e^s - 3)(e^s - 2)(e^s - 1)}$$

$$= \frac{e^s - 1}{s} \left(\frac{\frac{1}{2}}{e^s - 1} - \frac{1}{e^s - 2} + \frac{\frac{1}{2}}{e^s - 3} \right)$$

$$= \frac{1 - e^{-s}}{s} \left(\frac{\frac{1}{2}}{1 - e^{-s}} - \frac{1}{1 - 2e^{-s}} + \frac{\frac{1}{2}}{1 - 3e^{-s}} \right)$$

$$= \frac{1}{2} \mathcal{L}(1) - \mathcal{L}(2^{[t]}) + \frac{1}{2} \mathcal{L}(3^{[t]}).$$

Whence

$$a_n = \frac{1}{2} 3^n - \frac{1}{2}, \qquad n = 0, 1, 2, \ldots.$$

Linear difference equations involving derivatives of the function $y(t)$ can also be treated by the Laplace transform method.

Example 2.49. Solve the differential-difference equation

$$y''(t) - y(t - 1) = \delta(t), \qquad y(t) = y'(t) = 0, \ t \leq 0.$$

Similarly, as we saw in Example 2.46,

$$\mathcal{L}(y(t - 1)) = e^{-s} \mathcal{L}(y(t)).$$

Then transforming the original equation,

$$s^2 \mathcal{L}(y(t)) - e^{-s} \mathcal{L}(y(t)) = 1,$$

so that

$$\mathcal{L}(y(t)) = \frac{1}{s^2 - e^{-s}} = \frac{1}{s^2 \left(1 - \frac{e^{-s}}{s^2}\right)} \qquad (\mathcal{R}e(s) > 0)$$

$$= \sum_{n=0}^{\infty} \frac{e^{-ns}}{s^{2n+2}} \qquad (\mathcal{R}e(s) > 1). \qquad (2.69)$$

Observe that by (1.9) and (1.14)

$$\mathcal{L}^{-1}\left(\frac{e^{-ns}}{s^{2n+2}}\right) = u_n(t)\frac{(t - n)^{2n+1}}{(2n + 1)!}$$

$$= \begin{cases} \frac{(t-n)^{2n+1}}{(2n+1)!} & t \geq n \\ 0 & t < n. \end{cases} \qquad (2.70)$$

Hence by (2.69) and (2.70) and the linearity of $\mathcal{L}$,

$$\mathcal{L}(y(t)) = \mathcal{L}\left(\sum_{n=0}^{[t]} \frac{(t - n)^{2n+1}}{(2n + 1)!}\right)$$

and

$$y(t) = \sum_{n=0}^{[t]} \frac{(t - n)^{2n+1}}{(2n + 1)!}.$$

Exercises 2.9

1. Show that the function

$$f(t) = a^{[t]}, \qquad t > 0,$$

has exponential order on $[0, \infty)$.

2. (a) Show that the function

$$f(t) = [t], \qquad t > 0,$$

has Laplace transform

$$\mathcal{L}(f(t)) = \frac{1}{s(e^s - 1)} = \frac{e^{-s}}{s(1 - e^{-s})}.$$

(b) Show that the solution to

$$y(t+1) - y(t) = 1, \qquad y(t) = 0, \ t < 1,$$

is given by the function in part (a).

3. From the expression

$$\frac{e^{-s}}{s(1 - ae^{-s})} = \frac{e^{-s}}{s}(1 + ae^{-s} + a^2 e^{-2s} + \cdots),$$

deduce that

$$f(t) = \mathcal{L}^{-1}\left(\frac{e^{-s}}{s(1 - ae^{-s})}\right) = \begin{cases} \sum_{n=1}^{[t]} a^{n-1} & t \geq 1 \\ 0 & 0 < t < 1, \end{cases}$$

and for $a \neq 1$,

$$f(t) = \mathcal{L}^{-1}\left(\frac{e^{-s}}{s(1 - ae^{-s})}\right) = \frac{a^{[t]} - 1}{a - 1}.$$

4. Solve for a_n:

(a) $a_{n+2} - 7a_{n+1} + 12a_n = 0$ $\qquad a_0 = 0, \ a_1 = -1$
(b) $a_{n+2} - 7a_{n+1} + 12a_n = 2^n$ $\qquad a_0 = 0, \ a_1 = -1$
(c) $a_{n+1} + a_n = 1, \qquad a_0 = 0, \ a_1 = 1$
(d) $a_{n+2} - 2a_{n+1} + a_n = 0, \qquad a_0 = 0, \ a_1 = 1.$

5. The *Fibonacci* difference equation is given by

$$a_{n+2} = a_{n+1} + a_n, \qquad a_0 = 0, \ a_1 = 1.$$

Deduce that

$$a_n = \frac{1}{\sqrt{5}}\left[\left(\frac{1 + \sqrt{5}}{2}\right)^n - \left(\frac{1 - \sqrt{5}}{2}\right)^n\right], \qquad n = 0, 1, 2, \ldots.$$

6. Solve

(a) $y(t) + y(t-1) = e^t, \qquad y(t) = 0, \ t \leq 0$
(b) $y'(t) - y(t-1) = t, \qquad y(t) = 0, \ t \leq 0.$

7. Find a_n if

$$a_{n+2} - 5a_{n+1} + 6a_n = 4n + 2, \qquad a_0 = 0, \ a_1 = 1.$$

3

Complex Variable Theory

In this chapter we present an overview of the theory of complex variables, which is required for an understanding of the complex inversion formula discussed in Chapter 4. Along the way, we establish the analyticity of the Laplace transform (Theorem 3.1) and verify the differentiation formula (1.15) of Chapter 1 for a complex parameter (Theorem 3.3).

3.1 Complex Numbers

Complex numbers are ordered pairs of real numbers for which the rules of addition and multiplication are defined as follows: If $z = (x, y)$, $w = (u, v)$, then

$$z + w = (x + u, y + v),$$

$$zw = (xu - yv, xv + yu).$$

With these operations the complex numbers satisfy the same arithmetic properties as do the real numbers. The set of all complex numbers is denoted by $\mathbb{C}$.

We identify (real) a with $(a, 0)$ and denote $i = (0, 1)$, which is called the *imaginary number*. However, it is anything but imaginary in the common sense of the word. Observe that

$$z = (x, y) = (x, 0) + (0, y) = (x, 0) + (y, 0)(0, 1) = x + yi = x + iy.$$

It is these latter two forms that are typically employed in the theory of complex numbers, rather than the ordered-pair expression. But it is worth remembering that the complex number $x + iy$ is just the ordered-pair (x, y) and that $i = (0, 1)$. Moreover,

$$i^2 = (0, 1)(0, 1) = (-1, 0) = -1,$$

which can also be expressed as $i = \sqrt{-1}$.

The *real part* of $z = x+iy$, written $\mathcal{R}e(z)$, is the real number x, and the *imaginary part*, $\mathcal{I}m(z)$, is the real number y. The two complex numbers $z = x + iy$, $w = u + iv$ are equal if and only if $x = u$ and $y = v$, that is, their real and imaginary parts are the same.

The *modulus* (or *absolute value*) of z is $|z| = r = \sqrt{x^2 + y^2}$, and $|zw| = |z| \, |w|$. As with real numbers, the *triangle inequality* holds:

$$|z + w| \le |z| + |w|.$$

The *conjugate* of $z = x + iy$ is given by $\bar{z} = x - iy$ (Figure 3.1). Thus, $z\bar{z} = |z|^2$ and $\overline{z + w} = \bar{z} + \bar{w}$, $\overline{zw} = \bar{z}\,\bar{w}$.

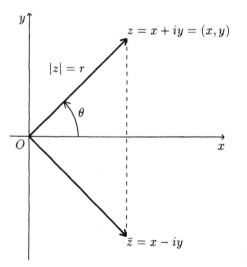

FIGURE 3.1

Complex numbers can be depicted in a plane known as the *complex plane* (Figure 3.1), also denoted by $\mathbb{C}$. The x-axis is called the *real axis* and the y-axis the *imaginary axis*. The complex number z can also be thought of as a vector that makes an angle θ with the real axis, and θ is called the *argument* of z, denoted by $\arg(z)$. Clearly,

$$x = r\cos\theta, \qquad y = r\sin\theta,$$

and

$$\tan\theta = \frac{y}{x}.$$

Thus, we have

$$z = x + iy = r(\cos\theta + i\sin\theta),$$

the *polar form* of z.

If $z = r(\cos\theta + i\sin\theta)$ and $w = R(\cos\varphi + i\sin\varphi)$, then

$$zw = rR\left[(\cos\theta\cos\varphi - \sin\theta\sin\varphi) + i(\sin\theta\cos\varphi + \cos\theta\sin\varphi)\right]$$

$$= rR\left[\cos(\theta + \varphi) + i\sin(\theta + \varphi)\right].$$

In other words, the arguments are additive under multiplication. Thus,

$$z^2 = r^2(\cos 2\theta + i\sin 2\theta),$$

and in general,

$$z^n = [r(\cos\theta + i\sin\theta)]^n = r^n(\cos n\theta + i\sin n\theta),$$

which is known as *De Moivre's theorem*.

The function e^z is defined by

$$e^z = e^{x+iy} = e^x(\cos y + i\sin y).$$

Setting $x = 0$ gives $e^{iy} = \cos y + i\sin y$, and the expression (*Euler's formula*)

$$e^{i\theta} = \cos\theta + i\sin\theta, \qquad 0 \le \theta < 2\pi,$$

represents any point on the *unit circle* $|z| = 1$ and leads to the

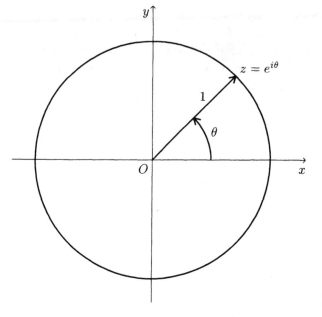

FIGURE 3.2

remarkable expression $e^{i\pi} = -1$ (Figure 3.2).

In general, therefore, any complex number $z = x + iy = r(\cos\theta + i\sin\theta)$ can be written as

$$z = r\,e^{i\theta}.$$

De Moivre's theorem now reads

$$z^n = (r\,e^{i\theta})^n = r^n e^{in\theta}.$$

If $r\,e^{i\theta} = z = w^n = R^n e^{in\varphi}$, then $w = z^{1/n}$ and

$$R = r^{\frac{1}{n}}, \qquad \varphi = \frac{\theta + 2k\pi}{n}, \qquad k = 0, \pm 1, \pm 2, \cdots.$$

Due to periodicity we need take only the values $k = 0, 1, \cdots, n - 1$ to obtain the n *distinct roots* of z:

$$z^{\frac{1}{n}} = r^{\frac{1}{n}} e^{i\left(\frac{\theta + 2k\pi}{n}\right)}.$$

For example, the four *fourth roots of unity* ($z = 1$) are given by

$$z^{\frac{1}{4}} = e^{i\left(\frac{0 + 2k\pi}{4}\right)}, \qquad k = 0, 1, 2, 3,$$

that is,

$$z_1 = 1, \qquad z_2 = e^{i\frac{\pi}{2}} = i, \qquad z_3 = e^{i\pi} = -1, \qquad z_4 = e^{i\frac{3\pi}{2}} = -i.$$

Exercises 3.1

1. If $z = 1 + 2i$, $w = 2 - i$, compute

    (a) $2z + 3w$ **(b)** $(3z)(2w)$

    (c) $\dfrac{3}{z} + \dfrac{2}{w}$.

2. Find the modulus, argument, real and imaginary parts of

    (a) $(1 + i)^3$ **(b)** $\dfrac{1 - i}{1 + i}$

    (c) $\dfrac{1}{(1 - i)^2}$ **(d)** $\dfrac{4 + 3i}{2 - i}$

    (e) $(1 + i)^{30}$.

3. Write the complex numbers in Question 2, parts (a) and (d), in polar form.

4. Show that if z is a complex number, then

    (a) $z + \bar{z} = 2\,Re(z)$
    (b) $z - \bar{z} = 2i\,Im(z)$
    (c) $|Re(z)| \le |z|$, $|Im(z)| \le |z|$.

5. Prove by mathematical induction that

$$|z_1 + z_2 + \cdots + z_n| \le |z_1| + |z_2| + \cdots + |z_n|, \qquad n \ge 2.$$

You may assume it is already valid for $n = 2$.

6. Show that

$$\left| \frac{z - a}{1 - \bar{a}z} \right| < 1$$

if $|z| < 1$ and $|a| < 1$. (Hint: $|w|^2 = w\bar{w}$.)

7. Determine the region in the z-plane denoted by

   (a) $|z - i| < 1$ **(b)** $1 \leq |z| \leq 2$

   (c) $\dfrac{\pi}{2} < \arg(z) < \dfrac{3\pi}{2}, \quad |z| < 1.$

8. Write in the form $x + iy$

   (a) $e^{i\frac{\pi}{4}}$ **(b)** $e^{2n\pi i}, \ n = 0, \pm 1, \pm 2, \ldots$

   (c) $e^{(2n-1)\pi i}, \ n = 0, \pm 1, \pm 2, \ldots$ **(d)** $e^{i\frac{5\pi}{3}}.$

9. Compute all the values of

   (a) $\sqrt[4]{-1}$ **(b)** $\sqrt[3]{i}$

   (c) $\sqrt[5]{1 + i}.$

3.2   Functions

A complex-valued function $w = f(z)$ of a complex variable assigns to each independent variable z one or more dependent variables w. If there is only one such value w, then the function $f(z)$ is termed *single-valued*; otherwise $f(z)$ is *multiple-valued*. Complex-valued functions are in general assumed to be single-valued unless otherwise stated. For $z = x + iy$ and $w = u + iv$, one can write

$$w = f(z) = u(x, y) + iv(x, y),$$

where $u = u(x, y)$, $v = v(x, y)$ are real-valued functions—the real and imaginary parts of $f(z)$.

    For example,

$$f(z) = z^2 = (x^2 - y^2) + 2ixy,$$

$$g(z) = e^z = e^x \cos y + i e^x \sin y,$$

$$h(z) = c \ = a + ib \qquad (a, b, c \ \text{constants})$$

are all (single-valued) complex functions.

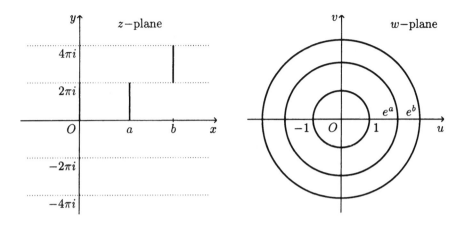

FIGURE 3.3

Complex functions are mappings from a domain in the z-plane to the range in the w-plane. For example, the exponential function $w = e^z = e^x e^{iy}$ maps the z-plane in such a way that each horizontal strip of width 2π is mapped onto the entire w-plane minus the origin (Figure 3.3). A vertical line at x in one of these strips maps to a circle of radius e^x in the w-plane. Note that when $x = 0$, then $e^{2n\pi i} = 1$, $n = 0, \pm1, \pm2, \ldots$, and $z_n = 2n\pi i$, $n = 0, \pm1, \pm2, \ldots$ are the only points that map to $w = 1$. Likewise $z_n = (2n - 1)\pi i$ are the only points that map to $w = -1$.

Functions related to the exponential function are as follows:

$$\sin z = \frac{e^{iz} - e^{-iz}}{2i}; \qquad \cos z = \frac{e^{iz} + e^{-iz}}{2};$$

$$\tan z = \frac{\sin z}{\cos z}; \qquad \cot z = \frac{\cos z}{\sin z};$$

$$\sinh z = \frac{e^z - e^{-z}}{2}; \qquad \cosh z = \frac{e^z + e^{-z}}{2};$$

$$\tanh z = \frac{\sinh z}{\cosh z} \left(z \neq (n - \tfrac{1}{2})\pi i\right); \qquad \coth z = \frac{\cosh z}{\sinh z} \left(z \neq n\pi i\right);$$

$$\operatorname{sech} z = \frac{1}{\cosh z} \left(z \neq (n - \tfrac{1}{2})\pi i\right); \qquad \operatorname{csch} z = \frac{1}{\sinh z} \left(z \neq n\pi i\right);$$

for $n = 0, \pm1, \pm2, \ldots$.

Some useful identities include

$$\sinh(z \pm w) = \sinh z \cosh w \pm \cosh z \sinh w,$$
$$\cosh(z \pm w) = \cosh z \cosh w \pm \sinh z \sinh w,$$

and, for $z = x + iy$,

$$\sinh z = \cos y \sinh x + i \sin y \cosh x,$$
$$\cosh z = \cos y \cosh x + i \sin y \sinh x.$$

An example of a multiple-valued complex function is the inverse of the exponential function, namely the logarithm function

$$\log z = \log |z| + i \arg(z) + 2n\pi i, \quad n = 0, \pm 1, \pm 2, \ldots, \quad 0 \le \arg(z) < 2\pi$$

which maps for each value of n the complex plane minus the origin onto the horizontal strips as in Figure 3.3 with the roles of the z- and w-planes reversed.

We call

$$w = \mathrm{Log}\, z = \log |z| + i \arg(z), \qquad 0 \le \arg(z) < 2\pi,$$

the *principal logarithm*. By removing the nonnegative real axis (a *branch cut*) from the domain, $\mathrm{Log}\, z$, as well as each of the *branches*,

$$\log z = \log |z| + i \arg(z) + 2n\pi i, \qquad n = 0, \pm 1, \pm 2, \ldots,$$

for each fixed n, becomes single-valued and analytic.

Another multiple-valued function is

$$w = z^{\frac{1}{n}} = r^{\frac{1}{n}} e^{i\left(\frac{\theta + 2k\pi}{n}\right)}, \qquad k = 0, 1, \ldots, n - 1,$$

which has n branches (one for each value of k) that are single-valued analytic for $0 < \theta < 2\pi$, $r > 0$, having again removed the nonnegative real axis. In particular, when $n = 2$, $w = \sqrt{z}$ has two branches:

$$w_1 = r^{\frac{1}{2}} e^{i\theta/2};$$

$$w_2 = r^{\frac{1}{2}} e^{i(\theta/2)+\pi} = -w_1.$$

We can even take a branch cut removing the nonpositive real axis so that w_1 and w_2 are (single-valued) analytic on $-\pi < \theta < \pi$, $r > 0$.

For the preceding multiple-valued functions, after one complete circuit of the origin in the z-plane, we find that the value of w shifts to another branch. Because of this property, $z = 0$ is termed a *branch point*. The branch point can be a point other than the origin; the function $w = \sqrt{z-1}$ has a branch point at $z = 1$.

Analytic Functions. The notions of limit and continuity are essentially the same for complex functions as for real functions. The only difference is that whenever $z \to z_0$ in a limit, the value of the limit should be independent of the direction of approach of z to z_0. Regarding the derivative, we say that a complex function $f(z)$ defined on a domain (connected open set) D is *differentiable at a point* $z_0 \in D$ if the limit

$$\frac{df}{dz}(z_0) = f'(z_0) = \lim_{z \to z_0} \frac{f(z) - f(z_0)}{z - z_0}$$

exists.

If $f(z)$ is differentiable at all points of some neighborhood $|z - z_0| < r$, then $f(z)$ is said to be *analytic (holomorphic)* at z_0. If $f(z)$ is analytic at each point of a domain D, then $f(z)$ is *analytic in D*. Since analytic functions are differentiable, they are continuous.

Differentiation of sums, products, and quotients of complex functions follow the same rules as for real functions. Moreover, there are the familiar formulas from real variables,

$$\frac{d}{dz} z^n = n z^{n-1},$$

$$\frac{d}{dz} e^z = e^z,$$

$$\frac{d}{dz} \text{Log} z = \frac{1}{z} \qquad \left(0 < \arg(z) < 2\pi\right),$$

and so forth.

Cauchy–Riemann Equations. For an analytic function $f(z) = u(x, y) + iv(x, y)$, the real and imaginary parts u and v cannot be arbitrary functions but must satisfy a special relationship known as the *Cauchy–Riemann equations*:

$$u_x = v_y; \qquad u_y = -v_x. \tag{3.1}$$

These arise from the fact that

$$f'(z_0) = u_x(z_0) + iv_x(z_0), \tag{3.2}$$

letting $z \to z_0$ along a line parallel to the real axis in computing the derivative, and

$$f'(z_0) = v_y(z_0) - i\, u_y(z_0), \tag{3.3}$$

letting $z \to z_0$ along a line parallel to the imaginary axis. Equating the real and imaginary parts of (3.2) and (3.3) gives (3.1).

One consequence of (3.1) is that a nonconstant analytic function $f = u + iv$ cannot have $v \equiv 0$ (that is, f is a real-valued function), for the Cauchy–Riemann equations would imply $u \equiv$ constant, a contradiction.

Equally important is the partial converse:

If $f(z) = u(x, y) + iv(x, y)$ is defined in a domain D and the partial derivatives u_x, u_y, v_x, v_y are continuous and satisfy the Cauchy–Riemann equations, then $f(z)$ is analytic in D.

Let us make use of this result to show that the Laplace transform is an analytic function.

Theorem 3.1. *Let $f(t)$ be piecewise continuous on $[0, \infty)$ and of exponential order α. Then*

$$F(s) = \mathcal{L}\big(f(t)\big)$$

is an analytic function in the domain $\mathcal{R}e(s) > \alpha$.

PROOF. For $s = x + iy$,

$$F(s) = \int_0^\infty e^{-st} f(t)\, dt = \int_0^\infty e^{-(x+iy)t} f(t)\, dt$$

$$= \int_0^\infty e^{-xt}(\cos yt - i\sin yt) f(t)\, dt$$

$$= \int_0^\infty (e^{-xt}\cos yt) f(t)\, dt + i\int_0^\infty (-e^{-xt}\sin yt) f(t)\, dt$$

$$= u(x, y) + iv(x, y).$$

Now consider

$$\left| \int_{t_0}^\infty \frac{\partial}{\partial x}(e^{-xt}\cos yt) f(t)\, dt \right| = \left| \int_{t_0}^\infty (-te^{-xt}\cos yt) f(t)\, dt \right|$$

$$\leq \int_{t_0}^\infty te^{-xt}|f(t)|\, dt$$

$$\leq M \int_{t_0}^{\infty} e^{-(x-\alpha-\delta)t} dt \qquad (\delta > 0)$$

$$\leq \frac{M}{x - \alpha - \delta} e^{-(x-\alpha-\delta)t_0},$$

where $\delta > 0$ can be chosen arbitrarily small. Then for $x \geq x_0 > \alpha$ (and hence $x \geq x_0 > \alpha + \delta$), the right-hand side can be made arbitrarily small by taking t_0 sufficiently large, implying that the integral $\int_0^{\infty} (\partial/\partial x)(e^{-xt} \cos yt) f(t) dt$ converges uniformly in $\mathcal{R}e(s) \geq x_0 > \alpha$.

Likewise, the integral $\int_0^{\infty} (\partial/\partial y)(-e^{-xt} \sin yt) f(t) dt$ converges uniformly in $\mathcal{R}e(s) \geq x_0 > \alpha$.

Because of this uniform convergence, and the absolute convergence of $\mathcal{L}(f)$, by Theorem A.12 we can differentiate under the integral sign, that is to say,

$$u_x = \int_0^{\infty} \frac{\partial}{\partial x} (e^{-xt} \cos yt) f(t) dt$$

$$= \int_0^{\infty} (-t\, e^{-xt} \cos yt) f(t) dt,$$

$$v_y = \int_0^{\infty} \frac{\partial}{\partial y} (-e^{-xt} \sin yt) f(t) dt$$

$$= \int_0^{\infty} (-t\, e^{-xt} \cos yt) f(t) dt,$$

and so $u_x = v_y$. In a similar fashion the reader is invited to show that $u_y = -v_x$. The continuity of these partial derivatives follows from Theorem A.2 applied to the function $g(t) = -t f(t)$ and taking the real and imaginary parts.

Thus, the Cauchy–Riemann conditions are satisfied and $F(s) = u(x, y) + iv(x, y)$ is an analytic function in the domain $\mathcal{R}e(s) > \alpha$, since any such point s will lie to the right of a vertical line at some $x_0 > \alpha$. □

Remark 3.2. In general, if $f \in L$, then $F(s)$ is analytic in some half-plane, $\mathcal{R}e(s) > x_0$ (cf. Doetsch [2], Theorem 6.1).

In view of the foregoing discussion, let us verify the following formula proved in Chapter 1 for a real parameter s (Theorem 1.34).

Theorem 3.3. *If f is piecewise continuous on* $[0, \infty)$ *of order* α *and has Laplace transform* $F(s)$*, then*

$$\frac{d^n}{ds^n} F(s) = \mathcal{L}\big((-1)^n t^n f(t)\big), \qquad n = 1, 2, 3, \dots \quad (\mathcal{R}e(s) > \alpha).$$

PROOF.   Writing $F(s) = u(x, y) + iv(x, y)$, where u, v are as in the preceding theorem, we have by (3.2)

$$F'(s) = u_x + iv_x$$

$$= \int_0^\infty (-t e^{-xt} \cos yt) f(t)\, dt + i \int_0^\infty (t e^{-xt} \sin yt) f(t)\, dt$$

$$= \int_0^\infty -t(e^{-xt} \cos yt - i e^{-xt} \sin yt) f(t)\, dt$$

$$= \int_0^\infty -t e^{-st} f(t)\, dt$$

$$= \mathcal{L}\big(-t f(t)\big).$$

Repeated application of this procedure gives the formula.   □

The real and imaginary parts of an analytic function $f = u+iv$ not only satisfy the Cauchy–Riemann equations, but taking the second partial derivatives [which we can do since $f(z)$ has derivatives of all orders; see formula (3.7)], we find that

$$\Delta u = u_{xx} + u_{yy} = 0, \tag{3.4}$$

and likewise for v. Since the second partial derivatives are also continuous, both u and v are *harmonic* functions, satisfying the *Laplace equation* (3.4), and Δ is the *Laplace operator*. Here v is called the *harmonic conjugate* of u and vice versa.

Exercises 3.2

1. Show that

(a) $e^z = 1$ if and only if $z = 2n\pi i$, $n = 0, \pm 1, \pm 2, \dots$
(b) $e^z = -1$ if and only if $z = (2n + 1)\pi i$, $n = 0, \pm 1, \pm 2, \dots$.

2. Compute

(a) $\mathrm{Log}(-1)$ (b) $\mathrm{Log}(-ei)$

(c) $\mathrm{Log}\left(\dfrac{1+i}{1-i}\right)$.

3. We define the *principal value* of z^w by

$$z^w = e^{w\,\mathrm{Log}\,z} = e^{w\left(\log|z|+i\arg(z)\right)}.$$

Find the principal value of

(a) $(i)^i$ (b) $(-1)^{\frac{1}{\pi}}$

(c) $(1+i)^{(1+i)}$.

4. Show that

(a) $\dfrac{d}{dz}\cos z = -\sin z$

(b) $\dfrac{d}{dz}\cosh z = \sinh z$

(c) $\dfrac{d}{dz}\tanh z = \operatorname{sech}^2 z$ $\left(z \neq (n-\tfrac{1}{2})\pi i\right)$.

5. Show that

(a) $\sinh(z \pm w) = \sinh z \cosh w \pm \cosh z \sinh w$

(b) $\cosh(z \pm w) = \cosh z \cosh w \pm \sinh z \sinh w$.

6. Show that for $z = x + iy$

(a) $\sinh z = \cos y \sinh x + i \sin y \cosh x$

(b) $\cosh z = \cos y \cosh x + i \sin y \sinh x$.

7. Prove that the functions $f(z) = \bar{z}$ and $g(z) = |z|$ are nowhere analytic.

8. (a) Show that the function

$$u(x, y) = x^3 - 3xy^2 + xy$$

is harmonic in $\mathbb{C}$.

(b) Show that the function

$$v(x, y) = 3x^2 y - y^3 - \frac{x^2}{2} + \frac{y^2}{2}$$

is harmonic in $\mathbb{C}$ and that $f = u + iv$ is analytic in $\mathbb{C}$, where u is given in part (a).

9. If $f(z)$ is analytic, show that

$$\frac{\partial^2}{\partial x^2} |f(z)|^2 + \frac{\partial^2}{\partial y^2} |f(z)|^2 = 4|f'(z)|^2.$$

10. Show that if $f = u + iv$ is an analytic function and $v \equiv$ constant, then $f \equiv$ constant.

3.3   Integration

Integrals of complex-valued functions are calculated over certain types of curves in the complex plane. A parametric representation of a continuous curve $C: z(t) = x(t) + iy(t)$, $\alpha \le t \le \beta$, is *smooth* if $z'(t)$ is continuous for $\alpha \le t \le \beta$ and $z'(t) \ne 0$ for $\alpha < t < \beta$.

A *contour* C is just a continuous curve that is piecewise smooth, that is, there is a subdivision $\alpha = t_0 < t_1 < \cdots < t_n = \beta$ and $z = z(t)$ is smooth on each subinterval $[t_{k-1}, t_k]$, $k = 1, \cdots, n$. The point $z(\alpha)$ is the *initial point*, $z(\beta)$ is the *terminal point*, and, if $z(\alpha) = z(\beta)$, C is *closed*. (See Figure 3.4.) If C does not cross itself, it is called *simple*. Simple, closed contours (see Figure 3.5) enjoy both properties and form an important class of curves. The *positive direction* along a simple, closed contour C keeps the interior of C to the left, that is, the curve is traversed counterclockwise. If ∞ is an interior point, however, the positive direction is clockwise, with ∞ on the left.

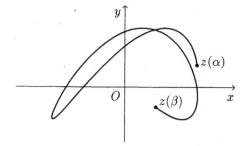

FIGURE 3.4

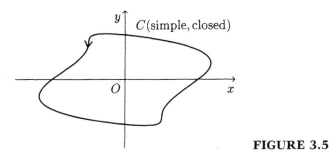

FIGURE 3.5

The reason for being so particular about the choice of curves is that for a continuous, complex-valued function $f(z)$ defined on a contour C, the (Riemann) integral of $f(z)$ over C can be defined as

$$\int_C f(z)\,dz = \int_\alpha^\beta f\big(z(t)\big)\,z'(t)\,dt \qquad (3.5)$$

since the right-hand integral exists. This is so because the integrand is piecewise continuous. In view of (3.5), many of the standard rules for integration carry over to the complex setting. One rule in particular is worth singling out:

$$-\int_C f(z)\,dz = \int_{-C} f(z)\,dz,$$

where $-C$ represents the contour C traversed in the opposite direction to that of C.

Furthermore, if $C_1, C_2, \ldots, C_n$ are disjoint contours, we define

$$\int_{C_1+C_2+\cdots+C_n} f(z)\,dz = \int_{C_1} f(z)\,dz + \int_{C_2} f(z)\,dz + \cdots + \int_{C_n} f(z)\,dz.$$

If $f(z)$ is continuous on contour C, then we can write

$$\left| \int_C f(z)\,dz \right| = \left| \int_\alpha^\beta f\big(z(t)\big)\,z'(t)\,dt \right| \le \int_\alpha^\beta \left| f\big(z(t)\big) \right|\, |z'(t)|\, dt$$

$$= \int_C |f(z)|\, |dz|,$$

where

$$\int_C |dz| = \int_\alpha^\beta |z'(t)|\, dt = \int_\alpha^\beta \sqrt{[x'(t)]^2 + [y'(t)]^2}\, dt$$

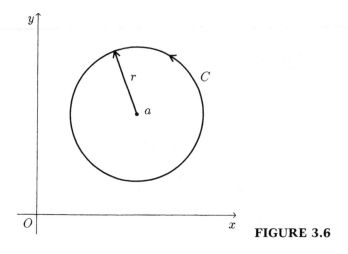

FIGURE 3.6

$$= \text{length of } C = L_C.$$

Thus, if $|f(z)| \leq M$ on C,

$$\left| \int_C f(z)\, dz \right| \leq \int_C |f(z)|\, |dz| \leq M L_C.$$

This type of estimate will be useful in the sequel.

Example 3.4. Let $C : z = a + r\, e^{it}$, $0 \leq t < 2\pi$ (Figure 3.6). Then $dz = ir\, e^{it} dt$ and

$$\int_C \frac{dz}{z-a} = \int_0^{2\pi} \frac{ir\, e^{it} dt}{r\, e^{it}} = 2\pi i.$$

Note that the function being integrated, $f(z) = 1/(z-a)$, is analytic in $\mathbb{C} - \{a\}$, but not at the point $z = a$.

In what follows, it is advantageous to consider our underlying domain in which we shall be integrating over closed contours, to "not contain any holes," unlike in the preceding example. To be more precise, we say that a domain D is *simply connected* if for any two continuous curves in D having the same initial and terminal points, either curve can be deformed in a continuous manner into the other while remaining entirely in D. The notion of a continuous deformation of one curve into another can be made mathematically precise, but that need not concern us here (cf. Ahlfors [1]).

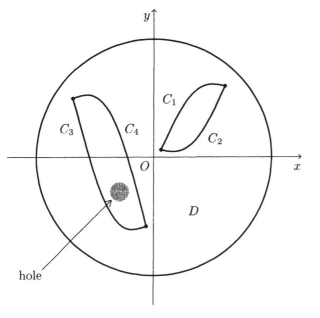

FIGURE 3.7

For example, the complex plane $\mathbb{C}$ is simply connected, as is any disk, and so is the domain $\mathbb{C} - \{\text{nonnegative real axis}\}$. On the other hand, $\mathbb{C} - \{a\}$ is not simply connected, nor is the annulus $A = \{z : 1 < |z| < 2\}$, nor the domain D in Figure 3.7.

This brings us to the cornerstone of complex variable theory:

Cauchy's Theorem. *Let $f(z)$ be analytic in a simply connected domain D. Then for any closed contour C in D,*

$$\int_C f(z)\,dz = 0.$$

One important consequence is that for any two points $z_1, z_2 \in D$ (simply connected) and $f(z)$ analytic,

$$\int_{z_1}^{z_2} f(z)\,dz$$

does not depend on the contour path of integration from z_1 to z_2, since the integral from z_1 to z_2 over contour C_1, followed by the integral from z_2 to z_1 over another contour C_2, gives by Cauchy's

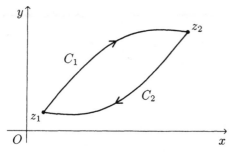

FIGURE 3.8

theorem

$$\int_{C_1+C_2} f(z)\, dz = 0$$

(Figure 3.8). Consequently,

$$\int_{C_1} f(z)\, dz = \int_{-C_2} f(z)\, dz,$$

and we say that the integral of $f(z)$ is *independent of path*.
This means that the integral

$$\int_{z_1}^{z_2} f(z)\, dz$$

can be evaluated in the manner of a real integral, that is, the integral
has the value $g(z_2) - g(z_1)$, where $g(z)$ is any antiderivative of $f(z)$,
namely, $g'(z) = f(z)$.

Example 3.5.   The integral

$$\int_{-i\pi}^{i\pi} \frac{dz}{z}$$

can be computed by taking any contour C lying in the left half-plane
that connects the points $-i\pi$ and $i\pi$ (Figure 3.9). Therefore,

$$\int_{-i\pi}^{i\pi} \frac{dz}{z} = \operatorname{Log} z \Big|_{-i\pi}^{i\pi} = \operatorname{Log}(i\pi) - \operatorname{Log}(-i\pi)$$

$$= \log|i\pi| + i \arg(i\pi) - \log|-i\pi| - i \arg(-i\pi)$$

$$= -i\pi.$$

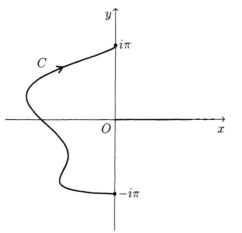

FIGURE 3.9

Cauchy Integral Formula. *Let $f(z)$ be analytic within and on a simple, closed contour C. If z_0 is any point interior to C, then*

$$f(z_0) = \frac{1}{2\pi i} \int_C \frac{f(z)\, dz}{z - z_0}, \tag{3.6}$$

taking the integration along C in the positive direction.

The hypothesis means that $f(z)$ is analytic on a slightly larger region containing C and its interior.

Furthermore, the nth derivative of $f(z)$ at $z = z_0$ is given by

$$f^{(n)}(z_0) = \frac{n!}{2\pi i} \int_C \frac{f(z)\, dz}{(z - z_0)^{n+1}}, \qquad n = 0, 1, 2, \ldots . \tag{3.7}$$

For $n = 0$ we have the Cauchy integral formula.

Example 3.6. Evaluate

$$\int_C \frac{e^z\, dz}{z^2 + 1},$$

where $C : |z| = 2$ is taken in the positive direction.

Taking a partial fraction decomposition and the Cauchy integral formula, we find

$$\int_C \frac{e^z}{z^2 + 1}\, dz = \frac{1}{2i} \int_C \frac{e^z}{z - i}\, dz - \frac{1}{2i} \int_C \frac{e^z}{z + i}\, dz$$

$$= \frac{1}{2i} 2\pi i \, e^i - \frac{1}{2i} 2\pi i e^{-i}$$

$$= \pi(e^i - e^{-i}).$$

If $f(z)$ is analytic within and on a circle $C : |z - z_0| = R$, and $M = \max_{|z|=R} |f(z)|$, then from (3.7) we have for $n = 0, 1, 2, \ldots$

$$|f^{(n)}(z_0)| \le \frac{n!}{2\pi} \int_C \frac{|f(z)| \, |dz|}{|z - z_0|^{n+1}} \le \frac{n!}{2\pi} \cdot \frac{M}{R^{n+1}} \cdot 2\pi R$$

$$= \frac{Mn!}{R^n}.$$

The condition $|f^{(n)}(z_0)| \le Mn!/R^n$ is known as *Cauchy's inequality*.

If M bounds all values of $|f(z)|$, $z \in \mathbb{C}$, namely $f(z)$ is bounded, as well as analytic in $\mathbb{C}$, then letting $R \to \infty$ in Cauchy's inequality with $n = 1$ gives $f'(z_0) = 0$. Since in this case z_0 is arbitrary, $f'(z) = 0$ for all $z \in \mathbb{C}$, implying $f \equiv$ constant in $\mathbb{C}$ by the Cauchy–Riemann equations. We have established the following result.

Liouville's Theorem. *Any bounded analytic function in $\mathbb{C}$ is constant.*

As an application, suppose that $f(z) = u(z) + iv(z)$ is analytic in $\mathbb{C}$ with $u(z) > 0$, $z \in \mathbb{C}$. Then the analytic function

$$F(z) = e^{-f(z)}$$

satisfies $|F(z)| = e^{-u(z)} < 1$ in $\mathbb{C}$, and Liouville's theorem implies $F(z) \equiv$ constant. Whence $f(z) \equiv$ constant.

Exercises 3.3

1. Compute the value of the following integrals over the given contour C traversed in the positive direction:

   (a) $\displaystyle\int_C \frac{dz}{z+1}$, $C : |z - 1| = 3$

   (b) $\displaystyle\int_C \bar{z} dz$, $C : |z| = 1$

(c) $\displaystyle \int_C z d\bar{z}$, C :

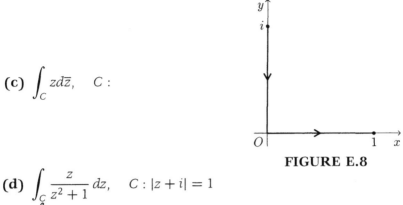

FIGURE E.8

(d) $\displaystyle \int_C \frac{z}{z^2 + 1} \, dz$, $C : |z + i| = 1$

(e) $\displaystyle \int_C e^z dz$, C is the perimeter of the square with vertices at
$z = 1 + i, z = -1 + i, z = -1 - i, z = 1 - i$

(f) $\displaystyle \int \frac{dz}{z^4 + 1}$, $C : |z| = 2$

(g) $\displaystyle \int_C \frac{(\cosh z + z^2)}{z(z^2 + 1)} \, dz$, $C : |z| = 2$

(h) $\displaystyle \int_C \frac{2z^4 + 3z^2 + 1}{(z - \pi i)^3} \, dz$, $C : |z| = 4$.

2. Compute the value of the following integrals:

(a) $\displaystyle \int_{-i\pi/2}^{i\pi/2} \frac{dz}{z}$

(b) $\displaystyle \int_{i\pi/2}^{i\pi} z e^z \, dz$.

3. Let C be the arc of the circle from $z = R$ to $z = -R$ that lies in the upper half-plane. Without evaluating the integral, show that

$$\left| \int_C \frac{e^{imz}}{z^2 + a^2} \, dz \right| \leq \frac{\pi R}{R^2 - a^2} \qquad (m > 0).$$

4. **(a)** Using the Cauchy integral formula, show that

$$f(a) - f(b) = \frac{a - b}{2\pi i} \int_{|z|=R} \frac{f(z)}{(z - a)(z - b)} \, dz,$$

for $f(z)$ analytic in $\mathbb{C}$, $|a| < R$, $|b| < R$.

(b) Use the result of part (a) to give another proof of Liouville's theorem.

5. Let $f(z)$ be analytic in the disk $|z| \leq 3$ and suppose that $|f(z)| \leq 5$ for all values of z on the circle $|z - 1| = 2$. Find an upper bound for $|f^{(4)}(0)|$.

6. Suppose that $f(z)$ is analytic in $\mathbb{C}$ and satisfies

$$|f(z)| \geq \frac{1}{10}, \qquad z \in \mathbb{C}.$$

Prove that $f(z) \equiv$ constant.

7. Suppose that $f(z)$ is analytic in $\mathbb{C}$ and satisfies

$$|f(z)| \leq |e^z|, \qquad z \in \mathbb{C}.$$

Show that $f(z) = ce^z$ for some constant c.

3.4 Power Series

A *power series* is an infinite series of the form

$$\sum_{n=0}^{\infty} a_n(z - z_0)^n = a_0 + a_1(z - z_0) + a_2(z - z_0)^2 + \cdots, \qquad (3.8)$$

where z is a complex variable and $z_0, a_0, a_1, \ldots$ are fixed complex numbers.

Every power series (3.8) has a *radius of convergence R*, with $0 \leq R \leq \infty$. If $R = 0$, then the series converges only for $z = z_0$. When $0 < R < \infty$, the series converges absolutely for $|z - z_0| < R$ and uniformly for $|z - z_0| \leq R_0 < R$. The series diverges for $|z - z_0| > R$. When $R = \infty$, the series converges for all $z \in \mathbb{C}$. The value of R is given by

$$R = \frac{1}{\varlimsup_{n \to \infty} \sqrt[n]{|a_n|}}$$

or by

$$R = \lim_{n \to \infty} \left| \frac{a_n}{a_{n+1}} \right|$$

whenever this limit exists.

Example 3.7.

(a) $\displaystyle\sum_{n=0}^{\infty} nz^n$ $R = \displaystyle\lim_{n\to\infty} \frac{n}{n+1} = 1.$

(b) $\displaystyle\sum_{n=0}^{\infty} \frac{z^n}{n!}$ $R = \displaystyle\lim_{n\to\infty} \frac{(n+1)!}{n!} = \infty.$

(c) $\displaystyle\sum_{n=0}^{\infty} n!z^n$ $R = \displaystyle\lim_{n\to\infty} \frac{n!}{(n+1)!} = 0.$

The circle $|z - z_0| = R$, when $0 < R < \infty$, is called the *circle of convergence*.

Two power series,

$$f(z) = \sum_{n=0}^{\infty} a_n(z - z_0)^n, \qquad g(z) = \sum_{n=0}^{\infty} b_n(z - z_0)^n,$$

that converge in a common disk $|z-z_0| < R$ can be added, subtracted, multiplied, and divided according to these rules:

- $f(z) \pm g(z) = \displaystyle\sum_{n=0}^{\infty}(a_n \pm b_n)(z - z_0)^n, \qquad |z - z_0| < R;$

- $f(z)\,g(z) = \displaystyle\sum_{n=0}^{\infty} c_n(z - z_0)^n, \qquad |z - z_0| < R,$

where

$$c_n = \sum_{k=0}^{n} a_k b_{n-k}, \qquad n = 0, 1, 2, \ldots;$$

- $\dfrac{f(z)}{g(z)} = \displaystyle\sum_{n=0}^{\infty} c_n(z - z_0)^n, \qquad |z - z_0| < r \le R,$

for $g(z) \neq 0$ in $|z - z_0| < r$, and c_n satisfies the recursive relation

$$c_n = \frac{a_n - c_0 b_n - c_1 b_{n-1} - \cdots - c_{n-1} b_1}{b_0} \qquad \left(g(z_0) = b_0 \neq 0\right).$$

A most significant feature of power series is that:

A power series represents an analytic function inside its circle of convergence.

Moreover, the converse is true:

If $f(z)$ is analytic in a disk $|z - z_0| < R$, then $f(z)$ has the Taylor series representation

$$f(z) = \sum_{n=0}^{\infty} \frac{f^{(n)}(z_0)}{n!} (z - z_0)^n \qquad (3.9)$$

at each point z in the disk.

The coefficients

$$a_n = \frac{f^{(n)}(z_0)}{n!}$$

are known as *Taylor coefficients*.

For example, the function $f(z) = \cosh z$ has the representation

$$\cosh z = \sum_{n=0}^{\infty} \frac{z^{2n}}{(2n)!} \qquad (z_0 = 0),$$

where $a_n = f^{(n)}(0)/n! = 1/n!$ (n even), $a_n = 0$ (n odd).

Suppose that $f(z)$ is not analytic in a complete disk but only in an annular region A bounded by two concentric circles $C_1 : |z - z_0| = R_1$ and $C_2 : |z - z_0| = R_2$, $0 < R_1 < R_2$ (Figure 3.10). We will assume that $f(z)$ is analytic on C_1 and C_2 as well, hence on a slightly larger

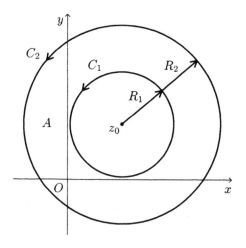

FIGURE 3.10

region. Then for each $z \in A$, we have the *Laurent series* representation

$$f(z) = \sum_{n=0}^{\infty} a_n(z - z_0)^n + \sum_{n=1}^{\infty} \frac{b_n}{(z - z_0)^n}, \qquad (3.10)$$

where

$$a_n = \frac{1}{2\pi i} \int_{C_2} \frac{f(\zeta)\, d\zeta}{(\zeta - z_0)^{n+1}}, \qquad n = 0, 1, 2, \ldots, \qquad (3.11)$$

$$b_n = \frac{1}{2\pi i} \int_{C_1} \frac{f(\zeta)\, d\zeta}{(\zeta - z_0)^{-n+1}}, \qquad n = 1, 2, 3, \ldots, \qquad (3.12)$$

and integration over C_1 and C_2 is in the positive direction.

This representation is a generalization of the Taylor series, for if $f(z)$ were analytic within and on C_2, then all the b_ns are zero by Cauchy's theorem since the integrands are analytic within and on C_1. Furthermore,

$$a_n = \frac{f^{(n)}(z_0)}{n!}, \qquad n = 0, 1, 2, \ldots$$

by (3.7).

Example 3.8. Let us determine the Laurent series representation of the function

$$f(z) = \frac{1}{(z - 1)(z + 2)}$$

in the annulus $1 < |z| < 2$.

First, by partial fractions we have

$$\frac{1}{(z - 1)(z + 2)} = \frac{1}{3(z - 1)} - \frac{1}{3(z + 2)}.$$

Since the *geometric series*

$$\sum_{n=0}^{\infty} \beta^n = 1 + \beta + \beta^2 + \cdots$$

converges for $|\beta| < 1$ to the value $1/(1 - \beta)$, and as we have in fact, $|1/z| < 1$ and $|z/2| < 1$, it follows that

$$\frac{1}{z - 1} = \frac{1}{z\left(1 - \frac{1}{z}\right)} = \frac{1}{z} \sum_{n=0}^{\infty} \frac{1}{z^n},$$

$$\frac{1}{z+2} = \frac{1}{2\left(1+\frac{z}{2}\right)} = \frac{1}{2}\sum_{n=0}^{\infty}(-1)^n\frac{z^n}{2^n}.$$

Hence

$$\frac{1}{(z-1)(z+2)} = \frac{1}{3}\sum_{n=0}^{\infty}\frac{1}{z^{n+1}} - \frac{1}{3}\sum_{n=0}^{\infty}\frac{(-1)^n z^n}{2^{n+1}}, \qquad 1 < |z| < 2.$$

This is the form of a Laurent series representation, and since the Laurent expansion is unique, we are done.

Singularities. A *singularity (singular point)* z_0 of a function $f(z)$ is a point at which $f(z)$ is not analytic, but any open disk about z_0, $|z - z_0| < R$, contains some point at which $f(z)$ is analytic. We say that z_0 is an *isolated singularity (isolated singular point)* if $f(z)$ is not analytic at z_0 but is analytic in a punctured disk, $0 < |z - z_0| < R$, of z_0.

For example,

$$f(z) = \frac{1}{(z-1)^2(z+2)}$$

has isolated singularities at $z = 1$, $z = -2$. On the other hand,

$$g(z) = \frac{1}{\sin\left(\frac{1}{z}\right)}$$

has isolated singularities at $z_n = 1/n\pi$, $n = \pm 1, \pm 2, \ldots$. The origin is also a singularity of $g(z)$ but not an isolated one since no punctured disk about $z = 0$ is free of singular points.

In this text we are concerned only with isolated singularities, of which there are three types.

If z_0 is an isolated singularity of $f(z)$, then we have the Laurent series representation (3.10)

$$f(z) = \sum_{n=1}^{\infty}\frac{b_n}{(z-z_0)^n} + \sum_{n=0}^{\infty}a_n(z-z_0)^n \qquad (3.13)$$

valid in some punctured disk $0 < |z - z_0| < R$.

(i) If $b_n = 0$ for all n, then for $z \neq z_0$ (3.13) becomes

$$f(z) = \sum_{n=0}^{\infty}a_n(z-z_0)^n.$$

Setting $f(z_0) = a_0$ makes $f(z)$ analytic at z_0, and z_0 is termed a *removable singularity*.

For example, the function

$$f(z) = \sin z / z = \sum_{n=0}^{\infty} (-1)^n z^{2n} / (2n+1)! \quad (z \neq 0)$$

has a removable singularity at $z = 0$ if we set $f(0) = 1$.

(ii) If all but finitely many b_ns are zero, say $b_n = 0$ for all $n > m \geq 1$ and $b_m \neq 0$, then

$$f(z) = \frac{b_1}{z - z_0} + \frac{b_2}{(z - z_0)^2} + \cdots + \frac{b_m}{(z - z_0)^m} + \sum_{n=0}^{\infty} a_n (z - z_0)^n.$$

$$(3.14)$$

In this case, we say that z_0 is a *pole of order m* of $f(z)$. If $m = 1$, then z_0 is a *simple pole* of $f(z)$.

As an illustration,

$$f(z) = \frac{e^z}{z^3} = \frac{1}{z^3} + \frac{1}{z^2} + \frac{1}{2!z} + \frac{1}{3!} + \cdots \quad (|z| > 0)$$

has a pole of order 3 at $z = 0$. From the Laurent representation (3.13), it is readily deduced that

$f(z)$ *has a pole of order m at z_0 if and only if*

$$f(z) = \frac{h(z)}{(z - z_0)^m},$$

where $h(z)$ is analytic at z_0, $h(z_0) \neq 0$.

Thus, the function

$$f(z) = \frac{1}{z^2 + 1} = \frac{1}{(z - i)(z + i)}$$

is seen to have simple poles at $z = \pm i$.

A function that is analytic except for having poles is called *meromorphic*.

(iii) If an infinite number of b_ns are not zero in (3.13), then z_0 is an *essential singularity* of $f(z)$.

The function

$$f(z) = e^{\frac{1}{z}} = 1 + \frac{1}{z} + \frac{1}{2!z^2} + \frac{1}{3!z^3} + \cdots + \frac{1}{n!z^n} + \cdots \qquad (|z| > 0)$$

has an essential singularity at $z = 0$.

Residues. For a function $f(z)$ with an isolated singularity at z_0 and Laurent series representation

$$f(z) = \sum_{n=1}^{\infty} \frac{b_n}{(z - z_0)^n} + \sum_{n=0}^{\infty} a_n(z - z_0)^n$$

in $0 < |z - z_0| < R$, the coefficient b_1, according to (3.12), is given by

$$b_1 = \frac{1}{2\pi i} \int_C f(\zeta) \, d\zeta$$

for $C : |z - z_0| = r < R$. This coefficient is very special because of its integral representation and is termed the *residue of* $f(z)$ *at* z_0, abbreviated by Res(z_0).

In the event $f(z)$ has a pole of order m at z_0, the algorithm

$$\text{Res}(z_0) = b_1 = \frac{1}{(m-1)!} \lim_{z \to z_0} \frac{d^{m-1}}{dz^{m-1}} [(z - z_0)^m f(z)] \qquad (3.15)$$

permits the easy determination of the residue. When z_0 is a simple pole (i.e., $m = 1$), we have

$$\text{Res}(z_0) = \lim_{z \to z_0} (z - z_0) f(z). \qquad (3.16)$$

This latter case can often be treated as follows. Suppose that

$$f(z) = \frac{p(z)}{q(z)},$$

where $p(z)$ and $q(z)$ are analytic at z_0, $p(z_0) \neq 0$, and $q(z)$ has a simple zero at z_0, whence $f(z)$ has a simple pole at z_0. Then $q(z) = (z - z_0) Q(z)$, $Q(z_0) \neq 0$, and $q'(z_0) = Q(z_0)$, implying

$$\text{Res}(z_0) = \lim_{z \to z_0} (z - z_0) \frac{p(z)}{q(z)} = \lim_{z \to z_0} \frac{p(z)}{\frac{q(z) - q(z_0)}{z - z_0}}$$

$$= \frac{p(z_0)}{q'(z_0)}. \qquad (3.17)$$

On the other hand, if $q(z_0) = 0$ and $q'(z_0) \neq 0$, then

$$q(z) = q'(z_0)(z - z_0) + \frac{q''(z_0)}{2!}(z - z_0)^2 + \cdots$$
$$= (z - z_0)Q(z),$$

where $Q(z_0) = q'(z_0) \neq 0$. That is, we have shown that z_0 is a simple zero of $q(z)$, hence a simple pole of $f(z)$.

Example 3.9. The function

$$f(z) = \frac{e^{(z^2)}}{(z - i)^3}$$

has a pole of order 3 at $z = i$. Therefore,

$$\text{Res}(i) = \frac{1}{2!} \lim_{z \to i} \frac{d^2}{dz^2}[(z - i)^3 f(z)] = \frac{1}{2} \lim_{z \to i} \frac{d^2}{dz^2} e^{(z^2)}$$
$$= \lim_{z \to i}[2z^2 e^{(z^2)} + e^{(z^2)}] = -\frac{1}{e}.$$

Example 3.10. For

$$f(z) = \frac{e^{az}}{\sinh z} = \frac{2e^{az}}{e^z - e^{-z}} = \frac{p(z)}{q(z)},$$

the poles of $f(z)$ are the zeros of $\sinh z$, that is, where $e^z = e^{-z}$, and so $e^{2z} = 1$, implying $z = z_n = n\pi i$, $n = 0, \pm 1, \pm 2, \ldots$. Since $p(z_n) \neq 0$ and

$$q'(z_n) = e^{n\pi i} + e^{-n\pi i} = (-1)^n \cdot 2 \neq 0,$$

the poles z_n of $f(z)$ are simple. Thus,

$$\text{Res}(z_n) = \frac{p(z_n)}{q'(z_n)} = (-1)^n e^{an\pi i}, \qquad n = 0, \pm 1, \pm 2, \ldots.$$

The reason for computing residues is the following:

Cauchy Residue Theorem. *Let $f(z)$ be analytic within and on a simple, closed contour C except at finitely many points $z_1, z_2, \ldots, z_n$ lying in the interior of C (Figure 3.11). Then*

$$\int_C f(z)\,dz = 2\pi i \sum_{i=1}^{n} \text{Res}(z_i),$$

where the integral is taken in the positive direction.

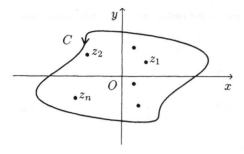

FIGURE 3.11

Example 3.11. Evaluate

$$\int_C \frac{e^{az}}{\cosh z}\, dz$$

for $C : |z| = 2$ in the positive direction.

Write

$$f(z) = \frac{e^{az}}{\cosh z} = \frac{p(z)}{q(z)}.$$

Then $\cosh z = (e^z + e^{-z})/2 = 0$ when $e^z = -e^{-z}$ (i.e., $e^{2z} = -1$), so that

$$z = z_n = \left(n - \tfrac{1}{2}\right)\pi i, \qquad n = 0, \pm 1, \pm 2, \ldots.$$

Now, $p(z_n) \neq 0$ with

$$q'(z_n) = \frac{e^{(n-\frac{1}{2})\pi i} - e^{-(n-\frac{1}{2})\pi i}}{2} = (-1)^{n+1} i \neq 0,$$

so that all the poles z_n of $f(z)$ are simple. Furthermore, only $z_1 = (\pi/2)i$ and $z_0 = (-\pi/2)i$ lie interior to C. Hence,

$$\text{Res}(z_1) = \frac{p(z_1)}{q'(z_1)} = \frac{e^{a\frac{\pi}{2} i}}{i},$$

$$\text{Res}(z_0) = \frac{p(z_0)}{q'(z_0)} = \frac{e^{-a\frac{\pi}{2} i}}{-i}.$$

Therefore,

$$\int_C \frac{e^{az}}{\cosh z}\, dz = 2\pi \left[e^{a\frac{\pi}{2} i} - e^{-a\frac{\pi}{2} i}\right]$$

$$= 4\pi i \sin\left(\frac{a\pi}{2}\right).$$

Exercises 3.4

1. Determine the radius of convergence of the following series:

(a) $\displaystyle\sum_{n=0}^{\infty} \frac{(-1)^n z^n}{n^2 + 1}$

(b) $\displaystyle\sum_{n=0}^{\infty} \frac{3^n z^n}{(2n + 1)!}$

(c) $\displaystyle\sum_{n=1}^{\infty} \frac{n(z - i)^n}{n + 1}$

(d) $\displaystyle\sum_{n=0}^{\infty} \frac{(-1)^n z^n}{2^{2n}(n!)^2}$.

2. Compute the Taylor series about $z_0 = 0$ for the following functions and determine the radius of convergence:

(a) $e^{(z^2)}$

(b) $\sinh z$

(c) $\dfrac{1}{1 - z}$

(d) $\log(1 + z)$.

3. Let $f(z)$ be analytic in $\mathbb{C}$ with Taylor series

$$f(z) = \sum_{n=0}^{\infty} a_n z^n.$$

If $|f(z)| \leq M(r)$ on $|z| = r$, show that

$$|a_n| \leq \frac{M(r)}{r^n}, \qquad n = 0, 1, 2, \ldots.$$

(Note: This is another version of Cauchy's inequality.)

4. Determine the nature of the singularities of the following functions:

(a) $\dfrac{1}{z(z^2 + 1)^2}$

(b) $\dfrac{e^{(z^2)}}{z^3}$

(c) $\sin \dfrac{1}{z}$

(d) $\dfrac{1 + \cos \pi z}{1 - z}$.

5. Write down the first three terms of the (Taylor/Laurent) series representation for each function:

(a) $\dfrac{z}{\sin z}$

(b) $\dfrac{1}{z \sinh z}$

(c) $\dfrac{\sinh \sqrt{z}}{\sqrt{z} \cosh \sqrt{z}}$.

6. Find the Laurent series expansion (in powers of z) of

$$f(z) = \frac{1}{z(z+1)(z-3)}$$

in the, regions:

(a) $0 < |z| < 1$ 　　　　　　　　　　　(b) $1 < |z| < 3$

(c) $|z| > 3$.

7. Find all the poles of the following functions and compute their residues:

(a) $\dfrac{z}{z^2 + a^2}$ 　　　　　　　　　　(b) $\dfrac{1}{z(1 + e^{az})}$

(c) $\dfrac{\sin z}{z^3}$.

8. Evaluate the following integrals over the contour C taken in the positive direction:

(a) $\displaystyle\int_C \frac{1 - e^z}{z^2}\, dz, \quad C : |z| = 1$

(b) $\displaystyle\int_C \frac{\cos z}{z^2 + 1}\, dz, \quad C : |z| = 2$

(c) $\displaystyle\int_C \cot z\, dz, \quad C : |z| = 4$

(d) $\displaystyle\int_C \frac{1 + e^z}{1 - e^z}\, dz, \quad C : |z| = 1$

(e) $\displaystyle\int_C \frac{dz}{z^2(z+2)(z-1)}, \quad C : |z| = 3$.

9. Evaluate the integral

$$\int_C \frac{e^{i\pi z}}{2z^2 - 5z - 3}, \quad C : |z| = 2$$

(taken in the positive direction) by using the

(i) Cauchy integral formula
(ii) method of residues.

3.5 Integrals of the Type $\int_{-\infty}^{\infty} f(x)\,dx$

Much of the complex variable theory presented thus far has been to enable us to evaluate real integrals of the form

$$\int_{-\infty}^{\infty} f(x)\,dx.$$

To this end, we transform the problem to a contour integral of the form

$$\int_{\Gamma_R} f(z)\,dz,$$

where Γ_R is the contour consisting of the segment $[-R, R]$ of the real axis together with the semicircle $C_R : z = R\,e^{i\theta}$, $0 \leq \theta \leq \pi$ (Figure 3.12).

Suppose that $f(z)$ is analytic in the complex place $\mathbb{C}$ except at finitely many poles, and designate by $z_1, z_2, \ldots, z_n$ the poles of $f(z)$ lying in the upper half-plane. By choosing R sufficiently large, z_1, $z_2, \ldots, z_n$ will all lie in the interior of Γ_R. Then by the Cauchy residue theorem,

$$2\pi i \sum_{i=1}^{n} \mathrm{Res}(z_i) = \int_{\Gamma_R} f(z)\,dz$$

$$= \int_{-R}^{R} f(x)\,dx + \int_{C_R} f(z)\,dz.$$

If we can demonstrate that

$$\lim_{R \to \infty} \int_{C_R} f(z)\,dz = 0,$$

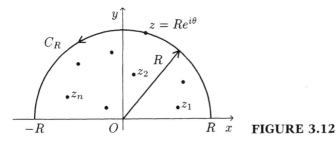

FIGURE 3.12

then we can deduce that the solution to our problem is given by

$$\int_{-\infty}^{\infty} f(x)\, dx = \lim_{R\to\infty} \int_{-R}^{R} f(x)\, dx = 2\pi i \sum_{i=1}^{n} \text{Res}(z_i). \qquad (3.18)$$

Example 3.12. Evaluate

$$\int_{-\infty}^{\infty} \frac{dx}{(x^2 + a^2)(x^2 + b^2)}, \qquad b > a > 0.$$

Let

$$f(z) = \frac{1}{(z^2 + a^2)(z^2 + b^2)},$$

and consider

$$\int_{\Gamma_R} f(z)\, dz,$$

where Γ_R is the contour in Figure 3.12. For R sufficiently large, the simple poles of $f(z)$ at the points ai and bi will be interior to Γ_R. Then by (3.16)

$$\text{Res}(ai) = \lim_{z\to ai} \frac{z - ai}{(z - ai)(z + ai)(z^2 + b^2)}$$

$$= \frac{1}{(2ai)(b^2 - a^2)},$$

$$\text{Res}(bi) = \lim_{z\to bi} \frac{z - bi}{(z^2 + a^2)(z - bi)(z + bi)}$$

$$= \frac{1}{(a^2 - b^2)(2bi)},$$

and

$$2\pi i\big(\text{Res}(ai) + \text{Res}(bi)\big) = \frac{\pi}{ab(a + b)}.$$

Moreover, on C_R, $z = R e^{i\theta}$, $|dz| = |iR e^{i\theta} d\theta| = R d\theta$, with $|z^2 + a^2| \geq |z|^2 - |a|^2 = R^2 - a^2$, $|z^2 + b^2| \geq R^2 - b^2$, and so

$$\left| \int_{C_R} \frac{dz}{(z^2 + a^2)(z^2 + b^2)} \right| \leq \int \frac{|dz|}{|z^2 + a^2|\,|z^2 + b^2|}$$

$$\leq \int_0^\pi \frac{R d\theta}{(R^2 - a^2)(R^2 - b^2)}$$

$$= \frac{\pi R}{(R^2 - a^2)(R^2 - b^2)} \to 0$$

as $R \to 0$.

Consequently, by the Cauchy residue theorem,

$$\frac{\pi}{ab(a+b)} = \int_{-R}^{R} \frac{dx}{(x^2 + a^2)(x^2 + b^2)} + \int_{C_R} \frac{dz}{(z^2 + a^2)(z^2 + b^2)},$$

and letting $R \to \infty$,

$$\int_{-\infty}^{\infty} \frac{dx}{(x^2 + a^2)(x^2 + b^2)} = \frac{\pi}{ab(a+b)}.$$

This example illustrates all the salient details, which will be exploited further in the next chapter.

Exercises 3.5

Use the methods of this section to verify the following integrals.

1. $\displaystyle\int_{-\infty}^{\infty} \frac{dx}{x^2 + x + 1} = \frac{2\pi}{\sqrt{3}}.$

2. $\displaystyle\int_0^{\infty} \frac{dx}{x^4 + x^2 + 1} = \frac{\pi\sqrt{3}}{6}.$

3. $\displaystyle\int_0^{\infty} \frac{x^2 dx}{1 + x^4} = \frac{\pi\sqrt{2}}{4}.$

4. $\displaystyle\int_0^{\infty} \frac{dx}{(x^2 + 1)(x^2 + 4)^2} = \frac{5\pi}{288}.$

5. $\displaystyle\int_0^{\infty} \frac{\cos x}{x^2 + a^2} dx = \frac{\pi e^{-a}}{2a}.$

[Hint: For this type of problem, consider the function

$$f(z) = \frac{e^{iz}}{z^2 + a^2},$$

and observe that on the x-axis

$$\mathcal{R}ef(z) = \mathcal{R}e\left(\frac{e^{ix}}{x^2 + a^2}\right) = \frac{\cos x}{x^2 + a^2}.$$

Now proceed as before.]

6. $\displaystyle\int_0^\infty \frac{x \sin mx}{x^2 + a^2}\, dx = \frac{\pi}{2}\, e^{-am}, \quad a > 0.$

[Hint: Consider the function

$$f(z) = \frac{ze^{imz}}{z^2 + a^2},$$

so that on the x-axis,

$$\mathcal{I}mf(z) = \frac{x \sin mx}{x^2 + a^2}.$$

Also, you will need the inequality $\sin\theta \geq 2\theta/\pi$, for $0 \leq \theta \leq \pi/2$.]

7. $\displaystyle\int_0^\infty \frac{dx}{x^n + 1} = \frac{\frac{\pi}{n}}{\sin\left(\frac{\pi}{n}\right)}, \quad n \geq 2.$

[Hint: Consider $\int_C \left(dz/(z^n + 1)\right)$ where C is the contour in Figure E.9.]

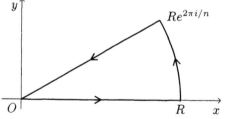

FIGURE E.9

4

CHAPTER

Complex Inversion Formula

The complex inversion formula is a very powerful technique for computing the inverse of a Laplace transform, $f(t) = \mathcal{L}^{-1}(F(s))$. The technique is based on the methods of contour integration discussed in Chapter 3 and requires that we consider our parameter s to be a complex variable.

For a continuous function f possessing a Laplace transform, let us extend f to $(-\infty, \infty)$ by taking $f(t) = 0$ for $t < 0$. Then for $s = x + iy$,

$$
\mathcal{L}\big(f(t)\big) = F(s) = \int_0^\infty e^{-st} f(t)\, dt
$$

$$
= \int_{-\infty}^\infty e^{-iyt}\big(e^{-xt} f(t)\big)\, dt
$$

$$
= F(x, y).
$$

In this form $F(x, y)$ represents the *Fourier transform* of the function $g(t) = e^{-xt} f(t)$. The Fourier transform is one of the most useful tools in mathematical analysis; its principal virtue is that it is readily inverted.

Towards this end, we assume that f is continuous on $[0, \infty)$, $f(t) = 0$ for $t < 0$, f has exponential order α, and f' is piecewise continuous on $[0, \infty)$. Then by Theorem 1.11, $\mathcal{L}\big(f(t)\big)$ converges absolutely for

151

$Re(s) = x > \alpha$, that is,

$$\int_0^\infty |e^{-st} f(t)| \, dt = \int_{-\infty}^\infty e^{-xt} |f(t)| \, dt < \infty, \qquad x > \alpha. \qquad (4.1)$$

This condition means that $g(t) = e^{-xt} f(t)$ is *absolutely integrable*, and we may thus invoke the Fourier inversion theorem (Theorem A.14), which asserts that $g(t)$ is given by the integral

$$g(t) = \frac{1}{2\pi} \int_{-\infty}^\infty e^{iyt} F(x, y) \, dy, \qquad t > 0.$$

This leads to the representation for f,

$$f(t) = \frac{1}{2\pi} \int_{-\infty}^\infty e^{xt} e^{iyt} F(x, y) \, dy, \qquad t > 0. \qquad (4.2)$$

Transforming (4.2) back to the variable $s = x + iy$, since $x > \alpha$ is fixed, we have $dy = (1/i) \, ds$ and so f is given by

$$f(t) = \frac{1}{2\pi i} \int_{x-i\infty}^{x+i\infty} e^{ts} F(s) \, ds = \lim_{y \to \infty} \frac{1}{2\pi i} \int_{x-iy}^{x+iy} e^{ts} F(s) \, ds. \qquad (4.3)$$

Here the integration is to be performed along a vertical line at $x > \alpha$ (Figure 4.1). The expression (4.3) is known as the *complex* (or *Fourier–Mellin*) *inversion formula*, and the vertical line at x is known as the *Bromwich line*. In order to calculate the integral in (4.3) and so determine the inverse of the Laplace transform $F(s)$, we employ the standard methods of contour integration discussed in Chapter 3.

To wit, take a semicircle C_R of radius R and center at the origin. Then for s on the *Bromwich contour* $\Gamma_R = ABCDEA$ of Figure 4.2,

$$\frac{1}{2\pi i} \int_{\Gamma_R} e^{ts} F(s) \, ds = \frac{1}{2\pi i} \int_{C_R} e^{ts} F(s) \, ds + \frac{1}{2\pi i} \int_{EA} e^{ts} F(s) \, ds. \qquad (4.4)$$

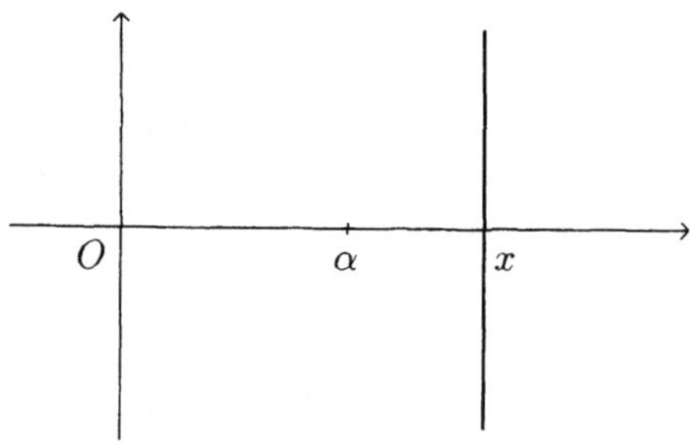

FIGURE 4.1

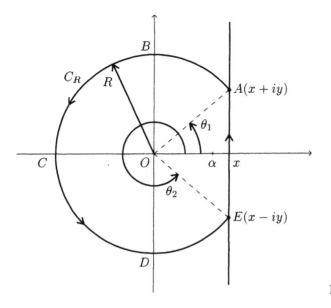

FIGURE 4.2

Since $F(s)$ is analytic for $\mathcal{R}e(s) = x > \alpha$, all the singularities of $F(s)$, such as they are, must lie to the left of the Bromwich line. For a preliminary investigation, let us *assume that $F(s)$ is analytic in $\mathcal{R}e(s) < \alpha$ except for having finitely many poles $z_1, z_2, \ldots, z_n$ there.* This is typical of the situation when, say

$$F(s) = \frac{P(s)}{Q(s)},$$

where $P(s)$ and $Q(s)$ are polynomials.

By taking R sufficiently large, we can guarantee that all the poles of $F(s)$ lie inside the contour Γ_R. Then by the Cauchy residue theorem,

$$\frac{1}{2\pi i} \int_{\Gamma_R} e^{ts} F(s)\, ds = \sum_{k=1}^{n} \operatorname{Res}(z_k), \tag{4.5}$$

where $\operatorname{Res}(z_k)$ is the residue of the function $e^{ts} F(s)$ at the pole $s = z_k$. Note that multiplying $F(s)$ by e^{ts} does not in any way affect the status of the poles z_k of $F(s)$ since $e^{ts} \neq 0$. Therefore, by (4.4) and (4.5),

$$\sum_{k=1}^{n} \operatorname{Res}(z_k) = \frac{1}{2\pi i} \int_{x-iy}^{x+iy} e^{ts} F(s)\, ds + \frac{1}{2\pi i} \int_{C_R} e^{ts} F(s)\, ds. \tag{4.6}$$

In what follows, we prove that when

$$\lim_{R \to \infty} \int_{C_R} e^{ts} F(s) \, ds = 0,$$

and then by letting $R \to \infty$ in (4.6), we obtain our desired conclusion:

$$f(t) = \lim_{y \to \infty} \frac{1}{2\pi i} \int_{x-iy}^{x+iy} e^{ts} F(s) \, ds = \sum_{k=1}^{n} \text{Res}(z_k). \qquad (4.7)$$

This formula permits the easy determination of the inverse function f. Let us then attend to the contour integral estimation.

$\int_{C_R} e^{ts} F(s) \, ds \to \Gamma$ **as** $R \to \infty$. An examination of the table of Laplace transforms (pp. 210–218) shows that most satisfy the growth restriction

$$|F(s)| \le \frac{M}{|s|^p}, \qquad (4.8)$$

for all sufficiently large values of $|s|$, and some $p > 0$.

For example, consider

$$F(s) = \frac{s}{s^2 - a^2} = \mathcal{L}^{-1}(\cosh at).$$

Then

$$|F(s)| \le \frac{|s|}{|s^2 - a^2|} \le \frac{|s|}{|s|^2 - |a|^2},$$

and for $|s| \ge 2|a|$, we have $|a|^2 \le |s|^2/4$, so that $|s|^2 - |a|^2 \ge 3|s|^2/4$, giving

$$|F(s)| \le \frac{4/3}{|s|} \qquad (|s| \ge 2|a|).$$

Observe that under the condition (4.8), $F(s) \to 0$ as $|s| \to \infty$.

Consider again the contour Γ_R as given in Figure 4.2. Any point s on the semicircle C_R is given by $s = R e^{i\theta}$, $\theta_1 \le \theta \le \theta_2$. Thus, $ds = iR e^{i\theta} d\theta$ and $|ds| = Rd\theta$.

Lemma 4.1. *For s on C_R, suppose that $F(s)$ satisfies*

$$|F(s)| \le \frac{M}{|s|^p}, \qquad \text{some } p > 0, \text{ all } R > R_0.$$

Then

$$\lim_{R \to \infty} \int_{C_R} e^{ts} F(s)\, ds = 0 \qquad (t > 0).$$

PROOF. For points $s = R\,e^{i\theta}$ on C_R, $|e^{ts}| = e^{tR\cos\theta}$. Therefore, for R sufficiently large so that all the poles of $F(s)$ are interior to Γ_R, $F(s)$ will be continuous on C_R with $|F(s)| \le M/R^p$ for all large R. Hence on the circular arc BCD,

$$\left| \int_{BCD} e^{st} F(s)\, ds \right| \le \int_{BCD} |e^{ts}|\, |F(s)|\, |ds|$$

$$\le \frac{M}{R^{p-1}} \int_{\frac{\pi}{2}}^{\frac{3\pi}{2}} e^{Rt\cos\theta}\, d\theta. \qquad (4.9)$$

At this stage substitute $\theta = \varphi + (\pi/2)$, which results in

$$\left| \int_{BCD} e^{st} F(s)\, ds \right| \le \frac{M}{R^{p-1}} \int_0^{\pi} e^{-Rt\sin\varphi}\, d\varphi$$

$$= \frac{2M}{R^{p-1}} \int_0^{\frac{\pi}{2}} e^{-Rt\sin\varphi}\, d\varphi, \qquad (4.10)$$

the latter equality being a consequence of $\sin\varphi$'s being symmetric about $\varphi = \pi/2$, for $0 \le \varphi \le \pi$.

In order to obtain a bound for the last integral, consider the graph of $y = \sin\varphi$, $0 \le \varphi \le \pi/2$ (Figure 4.3). The line from the origin to the point $(\pi/2, 1)$ has slope $m = 2/\pi < 1$, and thus the line $y = (2/\pi)\varphi$ lies under the curve $y = \sin\varphi$, that is,

$$\sin\varphi \ge \tfrac{2}{\pi}\varphi, \qquad 0 \le \varphi \le \tfrac{\pi}{2}.$$

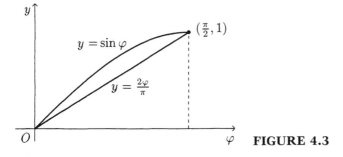

FIGURE 4.3

Consequently, (4.10) yields

$$\left| \int_{BCD} e^{ts} F(s) \, ds \right| \leq \frac{2M}{R^{p-1}} \int_0^{\frac{\pi}{2}} e^{-\frac{2Rt\varphi}{\pi}} \, d\varphi$$

$$= \frac{2M}{R^{p-1}} \frac{\pi}{-2Rt} \left[e^{-\frac{2Rt\varphi}{\pi}} \right]_0^{\frac{\pi}{2}}$$

$$= \frac{M\pi}{R^p t} (1 - e^{-Rt})$$

$$\rightarrow 0 \qquad \text{as} \quad R \rightarrow \infty.$$

Over the arc AB, we have $|e^{ts}| \leq e^{tx} = c$ for fixed $t > 0$, and the length of AB, $\ell(AB)$, remains bounded as $R \rightarrow \infty$, so that

$$\left| \int_{AB} e^{ts} F(s) \, ds \right| \leq \frac{cM\ell(AB)}{R^p} \rightarrow 0$$

as $R \rightarrow \infty$. Here we have taken x to be the value through which the Bromwich line passes, as in Figure 4.2.

Likewise,

$$\left| \int_{DE} e^{ts} F(s) \, ds \right| \rightarrow 0 \qquad \text{as} \quad R \rightarrow \infty.$$

As a consequence, we have our desired conclusion:

$$\lim_{R \rightarrow \infty} \int_{C_R} e^{ts} F(s) \, ds = 0. \qquad \square$$

Remarks 4.2.

i. We could have replaced the growth condition (4.8) with

$$|F(s)| \leq \varepsilon_R,$$

where $\varepsilon_R \rightarrow 0$ as $R \rightarrow \infty$, uniformly for s on C_R. For example,

$$F(s) = \frac{\log s}{s}$$

does satisfy this latter condition but not (4.8).

ii. If c_R is any subarc of C_R, say given by $\pi/2 \leq \theta_1' \leq \theta \leq \theta_2' \leq 3\pi/2$, then

$$\int_{\theta_1'}^{\theta_2'} e^{Rt \cos \theta} \, d\theta \leq \int_{\frac{\pi}{2}}^{\frac{3\pi}{2}} e^{Rt \cos \theta} \, d\theta$$

as the integrand is positive. Since the right-hand integral features in (4.9) and is ultimately bounded above by a quantity that tends to zero as $R \to 0$, we deduce that

$$\lim_{R \to \infty} \int_{C_R} e^{ts} F(s)\, ds = 0.$$

iii. Sometimes it is advantageous to use parabolas or other contours instead of semicircles (see Example 4.9).

Summarizing the result claimed in (4.7):

Theorem 4.3. *Suppose that f is continuous and f' piecewise continuous on $[0, \infty)$, with f of exponential order α on $[0, \infty)$. If $F(s) = \mathcal{L}(f(t))$, for $\mathcal{Re}(s) = x > \alpha$, also satisfies the growth condition*

$$|F(s)| \leq \frac{M}{|s|^p}, \qquad p > 0,$$

for all $|s|$ sufficiently large and some p (or condition (i) above), and if $F(s)$ is analytic in $\mathbb{C}$ except for finitely many poles at $z_1, z_2, \ldots, z_n$, then

$$f(t) = \frac{1}{2\pi i} \int_{x-i\infty}^{x+i\infty} e^{ts} F(s)\, ds = \sum_{k=1}^{n} \mathrm{Res}(z_k), \qquad (4.11)$$

where $\mathrm{Res}(z_k)$ is the residue of the function $e^{ts} F(s)$ at $s = z_k$.

In view of the properties of the inverse Fourier transform (Theorem A.14), we have the next result.

Corollary 4.4. *If f is only piecewise continuous on $[0, \infty)$, then the value returned by the complex inversion formula (4.11) is*

$$\frac{f(t^+) + f(t^-)}{2}$$

at any jump discontinuity $t > 0$.

Remark. The preceding theorem and corollary can be shown to hold under less restrictive conditions on f (see Doetsch [2], Theorem 24.4), so that functions such as $f(t) = 1/\sqrt{t}$ are not excluded by the inversion process. Essentially, the Laplace transform of f should converge absolutely and f should be of "bounded variation" in a neighborhood of the point $t > 0$ in question.

Example 4.5.

$$F(s) = \frac{1}{s(s-a)}.$$

Then $F(s)$ has a simple pole at $s = 0$ and $s = a$, and $|F(s)| \le M/|s|^2$ for all $|s|$ sufficiently large, say $|F(s)| \le 2/|s|^2$ if $|s| \ge 2|a|$. Moreover,

$$\mathrm{Res}(0) = \lim_{s \to 0} s e^{ts} F(s)$$

$$= \lim_{s \to 0} \frac{e^{ts}}{s-a} = -\frac{1}{a},$$

$$\mathrm{Res}(a) = \lim_{s \to a} (s-a) e^{ts} F(s)$$

$$= \lim_{s \to a} \frac{e^{ts}}{s} = \frac{e^{at}}{a}.$$

Whence

$$f(t) = \frac{1}{a}(e^{at} - 1).$$

Of course, $F(s)$ could have been inverted in this case using partial fractions or a convolution.

Example 4.6.

$$F(s) = \frac{1}{s(s^2 + a^2)^2} = \frac{1}{s(s-ai)^2(s+ai)^2}.$$

Then $F(s)$ has a simple pole at $s = 0$ and a pole of order 2 at $s = \pm ai$. Clearly, $|F(s)| \le M/|s|^5$ for all $|s|$ suitably large.

$$\mathrm{Res}(0) = \lim_{s \to 0} s e^{ts} F(s) = \lim_{s \to 0} \frac{e^{ts}}{(s^2 + a^2)^2} = \frac{1}{a^4}.$$

$$\mathrm{Res}(ai) = \lim_{s \to ai} \frac{d}{ds}\big((s-ai)^2 e^{ts} F(s)\big)$$

$$= \lim_{s \to ai} \frac{d}{ds}\left(\frac{e^{ts}}{s(s+ai)^2}\right)$$

$$= \frac{it}{4a^3} e^{iat} - \frac{e^{iat}}{2a^4}.$$

$$\text{Res}(-ai) = \lim_{s \to -ai} \frac{d}{ds} \left((s+ai)^2 e^{ts} F(s) \right)$$

$$= \lim_{s \to -ai} \frac{d}{ds} \left(\frac{e^{ts}}{s(s-ai)^2} \right)$$

$$= \frac{-it\, e^{-iat}}{4a^3} - \frac{e^{-iat}}{2a^4}.$$

Therefore,

$$\text{Res}(0) + \text{Res}(ai) + \text{Res}(-ai) = \frac{1}{a^4} + \frac{it}{4a^3}(e^{iat} - e^{-iat})$$

$$- \frac{1}{2a^4}(e^{iat} + e^{-iat})$$

$$= \frac{1}{a^4} \left(1 - \frac{a}{2} t \sin at - \cos at \right)$$

$$= f(t).$$

Example 4.7.

$$F(s) = \frac{P(s)}{Q(s)},$$

where $P(s)$ and $Q(s)$ are polynomials (having no common roots) of degree n and m, respectively, $m > n$, and $Q(s)$ has simple roots at $z_1, z_2, \ldots, z_m$. Then $F(s)$ has a simple pole at each $s = z_k$, and writing

$$F(s) = \frac{a_n s^n + a_{n-1} s^{n-1} + \cdots + a_0}{b_m s^m + b_{m-1} s^{m-1} + \cdots + b_0} \qquad (a_n, b_m \neq 0)$$

$$= \frac{a_n + \frac{a_{n-1}}{s} + \cdots + \frac{a_0}{s^n}}{s^{m-n} \left(b_m + \frac{b_{m-1}}{s} + \cdots + \frac{b_0}{s^m} \right)},$$

it is enough to observe that for $|s|$ suitably large,

$$\left| a_n + \frac{a_{n-1}}{s} + \cdots + \frac{a_0}{s^n} \right| \leq |a_n| + |a_{n-1}| + \cdots + |a_0| = c_1,$$

$$\left| b_m + \frac{b_{m-1}}{s} + \cdots + \frac{b_0}{s^m} \right| \geq |b_m| - \frac{|b_{m-1}|}{|s|} - \cdots - \frac{|b_0|}{|s|^m} \geq \frac{|b_m|}{2} = c_2,$$

and thus

$$|F(s)| \leq \frac{c_1/c_2}{|s|^{m-n}}.$$

Hence by (3.17),

$$\text{Res}(z_k) = \frac{e^{z_k t} P(z_k)}{Q'(z_k)}, \qquad k = 1, 2, \ldots, m,$$

and

$$f(t) = \sum_{k=1}^{m} \frac{P(z_k)}{Q'(z_k)} e^{z_k t}.$$

This is equivalent to the formulation given by (1.20).

Infinitely Many Poles. Suppose that $F(s)$ has infinitely many poles at $\{z_k\}_{k=1}^{\infty}$ all to the left of the line $\mathcal{R}e(s) = x_0 > 0$, and that

$$|z_1| \leq |z_2| \leq \cdots,$$

where $|z_k| \to \infty$ as $k \to \infty$. Choose a sequence of contours $\Gamma_n = C_n \cup [x_0 - iy_n, x_0 + iy_n]$ enclosing the first n poles $z_1, z_2, \ldots, z_n$ as in Figure 4.4. Then by the Cauchy residue theorem,

$$\frac{1}{2\pi i} \int_{\Gamma_n} e^{ts} F(s) \, ds = \sum_{k=1}^{n} \text{Res}(z_k),$$

where as before, $\text{Res}(z_k)$ is the residue of $e^{ts} F(s)$ at the pole $s = z_k$. Hence

$$\sum_{k=1}^{n} \text{Res}(z_k) = \frac{1}{2\pi i} \int_{x_0 - iy_n}^{x_0 + iy_n} e^{ts} F(s) \, ds + \frac{1}{2\pi i} \int_{C_n} e^{ts} F(s) \, ds.$$

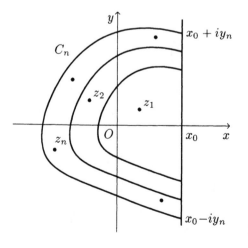

FIGURE 4.4

Once again, *if* it can be demonstrated that

$$\lim_{n\to\infty} \int_{C_n} e^{ts} F(s)\, ds = 0, \tag{4.12}$$

whereby $|y_n| \to \infty$, then we achieve the representation

$$f(t) = \frac{1}{2\pi i} \int_{x_0-i\infty}^{x_0+i\infty} e^{ts} F(s)\, ds = \sum_{k=1}^{\infty} \mathrm{Res}(z_k). \tag{4.13}$$

Example 4.8. Find

$$\mathcal{L}^{-1}\left(\frac{1}{s(1+e^{as})}\right), \quad a > 0.$$

The function

$$F(s) = \frac{1}{s(1+e^{as})}$$

has a simple pole at $s = 0$. Moreover, $1 + e^{as} = 0$ gives

$$e^{as} = -1 = e^{(2n-1)\pi i}, \quad n = 0, \pm 1, \pm 2, \ldots,$$

implying that

$$s_n = \left(\tfrac{2n-1}{a}\right)\pi i, \quad n = 0, \pm 1, \pm 2, \ldots,$$

are poles of $F(s)$.

For $G(s) = 1 + e^{as}$, $G'(s_n) = -a \neq 0$, which means that the poles s_n are simple. Furthermore,

$$\mathrm{Res}(0) = \lim_{s\to 0} s\, e^{ts} F(s) = \frac{1}{2},$$

$$\mathrm{Res}(s_n) = \frac{e^{ts_n}}{[s(1+e^{as})]'\Big|_{s=s_n}} = \frac{e^{ts_n}}{a\, s_n e^{as_n}}$$

$$= -\frac{e^{t\left(\frac{2n-1}{a}\right)\pi i}}{(2n-1)\,\pi i}.$$

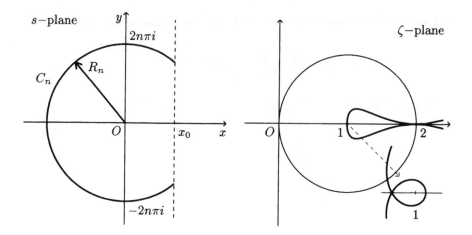

FIGURE 4.5

Consequently,

$$\text{sum of residues} = \frac{1}{2} - \sum_{n=-\infty}^{\infty} \frac{1}{(2n-1)\pi i} e^{t\left(\frac{2n-1}{a}\right)\pi i}$$

$$= \frac{1}{2} - \frac{2}{\pi} \sum_{n=1}^{\infty} \frac{1}{(2n-1)} \sin\left(\frac{2n-1}{a}\right)\pi t. \tag{4.14}$$

Finally, let C_n be the semicircle given by $s = R_n e^{i\theta}$, with $R_n = 2n\pi/a$. To make the subsequent reasoning simpler, let us take $a = 1$. Then the circles C_n cross the y-axis at the points $s = \pm 2n\pi i$. (See Figure 4.5.) We wish to consider what happens to the points s on C_n under the mapping $H(s) = 1 + e^s$.

(i) In the region $0 < x \le x_0$ and s on C_n, the image points $\zeta = H(s) = 1 + e^x e^{iy}$ all lie to the right and slightly below the point $\zeta = 2$, for y sufficiently close to $2n\pi$, that is, for n sufficiently large. (Notice that as n increases, the circles C_n flatten out so that $y = \mathcal{I}m(s)$ approaches $2n\pi$ from below.) Hence

$$|1 + e^s| \ge 2$$

for $0 \le x \le x_0$.

(ii) For s on C_n with $\mathcal{R}e(s) = x < 0$, the values of the function $H(s) = 1 + e^x e^{iy}$ lie inside the circle $|\zeta - 1| = 1$. As the value of $y = \mathcal{I}m(s)$ goes from $2n\pi$ down to $(2n-1)\pi$, the images spiral half a revolution with modulus

$$|1 + e^s| \geq 1 + e^{2n\pi \cos\varphi} \cos(2n\pi \sin\varphi) > b > 0,$$

for $x = 2n\pi \cos\varphi$, $y = 2n\pi \sin\varphi$.
As $y = \mathcal{I}m(s)$ goes from $(2n - 1)\pi$ down to $(2n - 2)\pi$, the images $H(s)$ spiral away from the origin half a revolution. For $y < 0$, it is the same story but spiraling outward.

Summarizing, the preceding shows that

$$|H(s)| = |1 + e^s| \geq c > 0$$

for some c, for all s on C_n, and likewise for $|1 + e^{as}|$. Consequently,

$$|F(s)| \leq \frac{c^{-1}}{|s|},$$

s on C_n, n sufficiently large. It follows that

$$\lim_{n\to\infty} \int_{C_n} e^{ts} F(s)\, ds = 0$$

in view of Lemma 4.1. The key here is that the contours C_n should straddle the poles.

We conclude that

$$f(t) = \mathcal{L}^{-1}\left(\frac{1}{s(1 + e^{as})}\right) = \frac{1}{2} - \frac{2}{\pi} \sum_{n=1}^{\infty} \frac{1}{(2n - 1)} \sin\left(\frac{2n - 1}{a}\right)\pi t,$$

as given by (4.14), at the points of continuity of f.

Remark. It should be observed that (4.14) is the Fourier series representation of the periodic square-wave function considered in Example 2.5. There we deduced that this function had Laplace transform $F(s) = 1/s(1 + e^{as})$. Note also that at the points of discontinuity, $t = na$, the series (4.14) gives the value $1/2$ (Figure 4.6).

Other useful inverses done in a similar fashion are $(0 < x < a)$

$$\mathcal{L}^{-1}\left(\frac{\sinh x\sqrt{s}}{s \sinh a\sqrt{s}}\right) = \frac{x}{a} + \frac{2}{\pi} \sum_{n=1}^{\infty} \frac{(-1)^n}{n} e^{-n^2\pi^2 t/a^2} \sin\frac{n\pi x}{a}, \qquad (4.15)$$

$$\mathcal{L}^{-1}\left(\frac{\cosh x\sqrt{s}}{s \cosh a\sqrt{s}}\right) = 1 + \frac{4}{\pi} \sum_{n=1}^{\infty} \frac{(-1)^n}{2n - 1} e^{-(2n-1)^2\pi^2 t/4a^2} \cos\left(\frac{2n - 1}{2a}\right)\pi x.$$

$$(4.16)$$

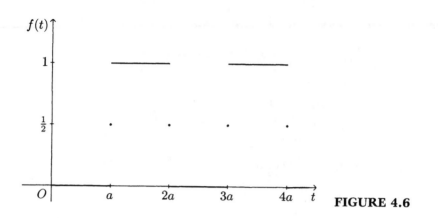

FIGURE 4.6

In the following example it is more appropriate to use parabolas instead of semicircles for the contours.

Example 4.9.

$$F(s) = \frac{\coth \sqrt{s}}{\sqrt{s}} = \frac{e^{\sqrt{s}} + e^{-\sqrt{s}}}{\sqrt{s}(e^{\sqrt{s}} - e^{-\sqrt{s}})}.$$

Setting

$$e^{\sqrt{s}} - e^{-\sqrt{s}} = 0$$

leads to

$$e^{2\sqrt{s}} = 1,$$

implying

$$2\sqrt{s} = 2n\pi i, \qquad n = \pm 1, \pm 2, \ldots,$$

and so

$$s_n = -n^2\pi^2, \qquad n = 1, 2, 3, \ldots$$

are simple poles of $F(s)$ since $(e^{\sqrt{s}} - e^{-\sqrt{s}})'|_{s=s_n} = (-1)^n/n\pi i \neq 0$. When $n = 0$, $F(s)$ also has a simple pole at $s_0 = 0$ because

$$\frac{e^{\sqrt{s}} + e^{-\sqrt{s}}}{\sqrt{s}(e^{\sqrt{s}} - e^{-\sqrt{s}})} = \frac{\left(1 + \sqrt{s} + \frac{s}{2!} + \cdots\right) + \left(1 - \sqrt{s} + \frac{s}{2!} - \cdots\right)}{\sqrt{s}\left[\left(1 + \sqrt{s} + \frac{s}{2!} + \cdots\right) - \left(1 - \sqrt{s} + \frac{s}{2!} - \cdots\right)\right]}$$

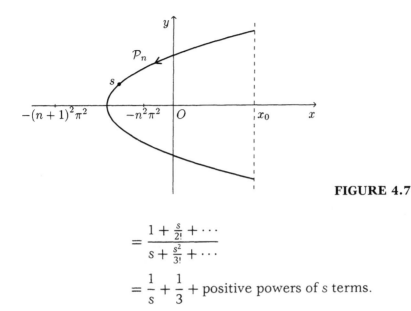

FIGURE 4.7

$$= \frac{1 + \frac{s}{2!} + \cdots}{s + \frac{s^2}{3!} + \cdots}$$

$$= \frac{1}{s} + \frac{1}{3} + \text{positive powers of } s \text{ terms.}$$

Let us consider the curve $\mathcal{P}_n$ in Figure 4.7 given by that part of the parabola

$$s = \left(\tau + i\left(n + \tfrac{1}{2}\right)\pi\right)^2$$

$$= \left(\tau^2 - \left(n + \tfrac{1}{2}\right)^2 \pi^2\right) + i2\tau\left(n + \tfrac{1}{2}\right)\pi$$

$$= x + iy,$$

for $x = \mathcal{R}e(s) < x_0$ ($x_0 > 0$) and τ a real parameter. Note that when $\tau = 0$,

$$x = -\left(n + \tfrac{1}{2}\right)^2 \pi^2, \qquad y = 0.$$

The advantage in taking this particular curve is that for s on $\mathcal{P}_n$,

$$\coth \sqrt{s} = \frac{e^{\tau + i\left(n + \frac{1}{2}\right)\pi} + e^{-\tau - i\left(n + \frac{1}{2}\right)\pi}}{e^{\tau + i\left(n + \frac{1}{2}\right)\pi} - e^{-\tau - i\left(n + \frac{1}{2}\right)\pi}}$$

$$= \frac{e^\tau - e^{-\tau}}{e^\tau + e^{-\tau}} = \tanh \tau.$$

Hence,

$$|F(s)| = \frac{|\tanh \tau|}{\left|\tau + i\left(n + \tfrac{1}{2}\right)\pi\right|}$$

$$\leq \frac{1}{\left(n + \frac{1}{2}\right)\pi}$$

$$= \varepsilon_n \to 0$$

as $n \to \infty$ (uniformly) for s on $\mathcal{P}_n$. By Exercises 4, Question 4, we conclude that

$$\lim_{n \to \infty} \int_{\mathcal{P}_n} e^{ts} F(s) \, ds = 0 \qquad (t > 0).$$

Regarding the residues, we have

$$\text{Res}(0) = 1,$$

$$\text{Res}(-n^2\pi^2) = \lim_{s \to -n^2\pi^2} (s + n^2\pi^2) e^{ts} \frac{\coth \sqrt{s}}{\sqrt{s}}$$

$$= \lim_{s \to -n^2\pi^2} \frac{e^{ts}}{\sqrt{s}} \cdot \lim_{s \to -n^2\pi^2} \frac{s + n^2\pi^2}{\tanh \sqrt{s}}$$

$$= \lim_{s \to -n^2\pi^2} \frac{e^{ts}}{\sqrt{s}} \cdot \lim_{s \to -n^2\pi^2} \frac{1}{\frac{1}{2\sqrt{s}} \operatorname{sech}^2 \sqrt{s}}$$

$$= 2e^{-n^2\pi^2 t}.$$

Finally,

$$f(t) = \mathcal{L}^{-1}\big(F(s)\big)$$

$$= \sum_{n=0}^{\infty} \text{Res}(-n^2\pi^2)$$

$$= 1 + 2 \sum_{n=1}^{\infty} e^{-n^2\pi^2 t} \qquad (t > 0).$$

What facilitated the preceding calculation of the inverse transform was the judicious choice of the parabolas $\mathcal{P}_n$. Herein lies the difficulty in determining the inverse of a meromorphic function $F(s)$ that has infinitely many poles. The curves C_n must straddle the poles, yet one must be able to demonstrate that $F(s) \to 0$ (uniformly) for s on C_n as $n \to \infty$. This task can be exceedingly difficult and may sometimes be impossible. It is tempting for practitioners of this technique, when $F(s)$ has infinitely many poles, not to verify (4.12) for suitable C_n. This leaves open the possibility that the resulting "inverse" function, $f(t)$, is incorrect.

Remark 4.10. There are many other variations where $F(s)$ involves the quotient of hyperbolic sines and hyperbolic cosines. See Doetsch [2], pp. 174–176, for further machinations involved with showing $\int_{C_n} e^{ts} F(s)\, ds \to 0$ as $n \to \infty$ via Lemma 4.1. Notwithstanding our preceding caveat, we will assume in Chapter 5 that $\int_{C_n} e^{ts} F(s)\, ds \to 0$ as $n \to \infty$ where required.

Branch Point. Consider the function

$$F(s) = \frac{1}{\sqrt{s}},$$

which has a branch point at $s = 0$. Although the inverse Laplace transform of $F(s)$ has already been considered in (2.5), it is instructive to apply the methods of the complex inversion formula in this case.

Consider the contour $C_R = ABCDEFA$, where AB and EF are arcs of a circle of radius R centered at O and CD is an arc γ_r of a circle of radius r also with center O (Figure 4.8).

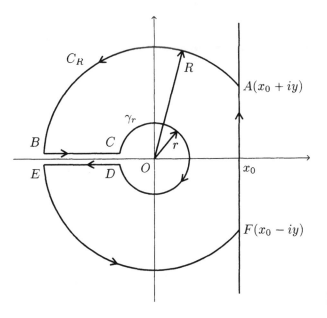

FIGURE 4.8

For $w = \sqrt{s}$ we take a branch cut along the nonpositive real axis with $-\pi < \theta < \pi$ and consider a (single-valued) analytic branch of w. Then $F(s) = 1/\sqrt{s}$ is analytic within and on C_R so that by Cauchy's theorem

$$\int_{C_R} \frac{e^{ts}}{\sqrt{s}}\, ds = 0.$$

Whence

$$0 = \frac{1}{2\pi i} \int_{x_0-iy}^{x_0+iy} \frac{e^{ts}}{\sqrt{s}}\, ds + \frac{1}{2\pi i} \int_{AB} \frac{e^{ts}}{\sqrt{s}}\, ds + \frac{1}{2\pi i} \int_{BC} \frac{e^{ts}}{\sqrt{s}}\, ds$$

$$+ \frac{1}{2\pi i} \int_{\gamma_r} \frac{e^{ts}}{\sqrt{s}}\, ds + \frac{1}{2\pi i} \int_{DE} \frac{e^{ts}}{\sqrt{s}}\, ds + \frac{1}{2\pi i} \int_{EF} \frac{e^{ts}}{\sqrt{s}}\, ds. \quad (4.17)$$

For $s = R e^{i\theta}$ lying on the two arcs AB and EF, we have

$$|F(s)| = \frac{1}{|s|^{\frac{1}{2}}},$$

so that by Remark 4.2, part (ii), coupled with the argument used in the proof of Lemma 4.1 to treat the portions of these arcs from A to $x = 0$ and from $x = 0$ to F, we conclude that

$$\lim_{R \to \infty} \int_{AB} \frac{e^{ts}}{\sqrt{s}}\, ds = \lim_{R \to \infty} \int_{EF} \frac{e^{ts}}{\sqrt{s}}\, ds = 0.$$

For $s = r e^{i\theta}$ on γ_r,

$$\left| \int_{\gamma_r} \frac{e^{ts}}{\sqrt{s}}\, ds \right| \leq \int_{\pi}^{-\pi} \frac{e^{tr\cos\theta}}{r^{\frac{1}{2}}}\, r d\theta$$

$$= r^{\frac{1}{2}} \int_{\pi}^{-\pi} e^{tr\cos\theta} d\theta \to 0$$

as $r \to 0$ since the integrand is bounded.

Finally, we need to consider the integrals along BC and DE. The values of these integrals converge to the values of the corresponding integrals when BC and DE are the upper and lower edges, respectively, of the cut along the negative x-axis. So it suffices to compute the latter. For s on BC, $s = x e^{i\pi}$, $\sqrt{s} = \sqrt{x} e^{i\pi/2} = i\sqrt{x}$, and when s

goes from $-R$ to $-r$, x goes from R to r. Hence

$$\int_{BC} \frac{e^{ts}}{\sqrt{s}} \, ds = \int_{-R}^{-r} \frac{e^{ts}}{\sqrt{s}} \, ds = -\int_R^r \frac{e^{-tx}}{i\sqrt{x}} \, dx$$

$$= \frac{1}{i} \int_r^R \frac{e^{-tx}}{\sqrt{x}} \, dx. \tag{4.18}$$

Along DE, $s = x e^{-i\pi}$, $\sqrt{s} = \sqrt{x} e^{-\frac{i\pi}{2}} = -i\sqrt{x}$, and

$$\int_{DE} \frac{e^{ts}}{\sqrt{s}} \, ds = \int_{-r}^{-R} \frac{e^{ts}}{\sqrt{s}} \, ds = -\int_r^R \frac{e^{-tx}}{-i\sqrt{x}} \, dx$$

$$= \frac{1}{i} \int_r^R \frac{e^{-tx}}{\sqrt{x}} \, dx. \tag{4.19}$$

Combining (4.18) and (4.19) after multiplying each by $1/2\pi i$ gives

$$\frac{1}{2\pi i} \int_{BC} \frac{e^{ts}}{\sqrt{s}} \, ds + \frac{1}{2\pi i} \int_{DE} \frac{e^{ts}}{\sqrt{s}} \, ds = -\frac{1}{\pi} \int_r^R \frac{e^{-tx}}{\sqrt{x}} \, dx.$$

Letting $R \to \infty$ and $r \to 0$ in (4.17) yields

$$0 = \frac{1}{2\pi i} \int_{x_0 - i\infty}^{x_0 + i\infty} \frac{e^{ts}}{\sqrt{s}} \, ds - \frac{1}{\pi} \int_0^\infty \frac{e^{-tx}}{\sqrt{x}} \, dx;$$

in other words,

$$f(t) = \mathcal{L}^{-1}\left(\frac{1}{\sqrt{s}}\right) = \frac{1}{2\pi i} \int_{x_0 - i\infty}^{x_0 + i\infty} \frac{e^{ts}}{\sqrt{s}} \, ds$$

$$= \frac{1}{\pi} \int_0^\infty \frac{e^{-tx}}{\sqrt{x}} \, dx.$$

To compute this latter integral, observe that by Example 2.1

$$\Gamma(\tfrac{1}{2}) = \int_0^\infty \frac{e^{-u}}{\sqrt{u}} \, du = \sqrt{\pi}.$$

Setting $u = tx$, $du = tdx$ and

$$\sqrt{\pi} = \int_0^\infty \frac{e^{-tx}}{\sqrt{t}\sqrt{x}} \, tdx = \sqrt{t} \int_0^\infty \frac{e^{-tx}}{\sqrt{x}} \, dx.$$

Therefore,

$$f(t) = \frac{1}{\pi}\left(\frac{\sqrt{\pi}}{\sqrt{t}}\right) = \frac{1}{\sqrt{\pi t}},$$

in accordance with (2.5).

Another useful example involving a branch point that arises in the solution of certain partial differential equations (see Section 5.1) is the determination of

$$\mathcal{L}^{-1}\left(\frac{e^{-a\sqrt{s}}}{s}\right), \qquad a > 0.$$

As in the preceding example, $s = 0$ is a branch point. Thus we can use the same contour (Figure 4.8) and approach in applying the complex inversion formula.

For $w = \sqrt{s}$ we take a branch cut along the nonpositive real axis and consider the (single-valued) analytic branch $w_1 = \sqrt{|s|}e^{i\theta/2}$ with positive real part.

Again, $F(s) = e^{-a\sqrt{s}}/s$ is analytic within and on C_R so that

$$\int_{C_R} \frac{e^{ts}e^{-a\sqrt{s}}}{s}\,ds = \int_{C_R} \frac{e^{ts-a\sqrt{s}}}{s}\,ds = 0.$$

Thus,

$$0 = \frac{1}{2\pi i}\int_{x_0-iy}^{x_0+iy} \frac{e^{ts-a\sqrt{s}}}{s}\,ds + \frac{1}{2\pi i}\int_{AB} \frac{e^{ts-a\sqrt{s}}}{s}\,ds$$

$$+ \frac{1}{2\pi i}\int_{BC} \frac{e^{ts-a\sqrt{s}}}{s}\,ds + \frac{1}{2\pi i}\int_{\gamma_r} \frac{e^{ts-a\sqrt{s}}}{s}\,ds$$

$$+ \frac{1}{2\pi i}\int_{DE} \frac{e^{ts-a\sqrt{s}}}{s}\,ds + \frac{1}{2\pi i}\int_{EF} \frac{e^{ts-a\sqrt{s}}}{s}\,ds. \qquad (4.20)$$

For $s = Re^{i\theta}$ on the two circular arcs AB and EF, $w_1 = \sqrt{s} = \sqrt{R}e^{i\theta/2}$ and

$$|F(s)| = \left|\frac{e^{-a\sqrt{s}}}{s}\right| = \frac{e^{-a\sqrt{R}\cos\theta/2}}{|s|} < \frac{1}{|s|},$$

and so as in the preceding example,

$$\lim_{R\to\infty}\int_{AB} \frac{e^{ts-a\sqrt{s}}}{s}\,ds = \lim_{R\to\infty}\int_{EF} \frac{e^{ts-a\sqrt{s}}}{s}\,ds = 0.$$

For s on the line segments BC and DE, again we take them to be the respective upper and lower edges of the cut along the negative axis. If s lies on BC, then $s = xe^{i\pi}$, $\sqrt{s} = i\sqrt{x}$, and when s goes from $-R$ to $-r$, x goes from R to r. Therefore,

$$\int_{BC} \frac{e^{ts-a\sqrt{s}}}{s} \, ds = \int_{-R}^{-r} \frac{e^{ts-a\sqrt{s}}}{s} \, ds = \int_{R}^{r} \frac{e^{-tx-ai\sqrt{x}}}{x} \, dx. \qquad (4.21)$$

Along DE, similarly $s = xe^{-i\pi}$, $\sqrt{s} = -i\sqrt{x}$ implying

$$\int_{DE} \frac{e^{ts-a\sqrt{s}}}{s} \, ds = \int_{r}^{R} \frac{e^{-tx+ai\sqrt{x}}}{x} \, dx. \qquad (4.22)$$

Combining (4.21) and (4.22) after multiplying each by $1/2\pi i$ yields

$$\frac{1}{2\pi i} \int_{r}^{R} \frac{e^{-tx}(e^{ai\sqrt{x}} - e^{-ai\sqrt{x}})}{x} \, dx = \frac{1}{\pi} \int_{r}^{R} \frac{e^{-tx} \sin a\sqrt{x}}{x} \, dx. \qquad (4.23)$$

Letting $r \to 0$ and $R \to \infty$ in (4.23), we obtain the integral

$$\frac{1}{\pi} \int_{0}^{\infty} \frac{e^{-tx} \sin a\sqrt{x}}{x} \, dx. \qquad (4.24)$$

In Section 2.7 we introduced the error function

$$\text{erf}(t) = \frac{2}{\sqrt{\pi}} \int_{0}^{t} e^{-x^2} \, dx.$$

It can be shown that the integral in (4.24) can be written in terms of the error function (see Theorem A.13), that is,

$$\frac{1}{\pi} \int_{0}^{\infty} \frac{e^{-tx} \sin a\sqrt{x}}{x} \, dx = \text{erf}\left(\frac{a}{2\sqrt{t}}\right);$$

we shall use use latter expression.

Finally, for $s = re^{i\theta}$ on γ_r, we can take the integration from π to $-\pi$,

$$\frac{1}{2\pi i} \int_{\gamma_r} \frac{e^{ts-a\sqrt{s}}}{s} \, ds = \frac{1}{2\pi i} \int_{\pi}^{-\pi} \frac{e^{tre^{i\theta}-a\sqrt{r}e^{i\theta/2}} ire^{i\theta} \, d\theta}{re^{i\theta}}$$

$$= -\frac{1}{2\pi} \int_{-\pi}^{\pi} e^{tre^{i\theta}-a\sqrt{r}e^{i\theta/2}} \, d\theta \to -\frac{1}{2\pi} \int_{-\pi}^{\pi} d\theta$$

$$= -1$$

for $r \to 0$. We have used here the uniform continuity of the integrand to pass the limit inside the integral.*

Whence, letting $r \to 0$, $R \to \infty$ in (4.20) gives

$$0 = \frac{1}{2\pi i} \int_{x_0 - i\infty}^{x_0 + i\infty} \frac{e^{ts - a\sqrt{s}}}{s} \, ds + \mathrm{erf}\left(\frac{a}{2\sqrt{t}}\right) - 1; \tag{4.25}$$

in other words,

$$f(t) = \mathcal{L}^{-1}\left(\frac{e^{-a\sqrt{s}}}{s}\right) = \frac{1}{2\pi i} \int_{x_0 - i\infty}^{x_0 + i\infty} \frac{e^{ts - a\sqrt{s}}}{s} \, ds$$

$$= 1 - \mathrm{erf}\left(\frac{a}{2\sqrt{t}}\right). \tag{4.26}$$

The function

$$\mathrm{erfc}(t) = 1 - \mathrm{erf}(t)$$

is called the *complementary error function*, and so we have by (4.26)

$$\mathcal{L}^{-1}\left(\frac{e^{-a\sqrt{s}}}{s}\right) = \mathrm{erfc}\left(\frac{a}{2\sqrt{t}}\right). \tag{4.27}$$

---

*Since for fixed a, t,

$$f(r, \theta) = e^{tre^{i\theta} - a\sqrt{r}e^{i\theta/2}}$$

is continuous on the closed rectangle $0 \le r \le r_0$, $-\pi \le \theta \le \pi$, it is uniformly continuous there. Hence for $\varepsilon > 0$, there exists $\delta = \delta(\varepsilon) > 0$ such that

$$|f(r, \theta) - f(r', \theta')| < \varepsilon \quad \text{whenever} \quad |(r, \theta) - (r', \theta')| < \delta.$$

In particular,

$$|f(r, \theta) - f(0, \theta)| < \varepsilon \quad \text{whenever} \quad 0 < r < \delta.$$

Then

$$\left| \int_{-\pi}^{\pi} f(r, \theta) \, d\theta - \int_{-\pi}^{\pi} f(0, \theta) \, d\theta \right| \le \int_{-\pi}^{\pi} |f(r, \theta) - f(0, \theta)| \, d\theta < 2\pi\varepsilon,$$

that is,

$$\lim_{r \to 0} \int_{-\pi}^{\pi} f(r, \theta) \, d\theta = \int_{-\pi}^{\pi} \lim_{r \to 0} f(r, \theta) \, d\theta = \int_{-\pi}^{\pi} f(0, \theta) \, d\theta.$$

Exercises

1. Using the method of residues (Theorem 4.3), determine the function $f(t)$ if the Laplace transform $F(s)$ is given by

   (a) $\dfrac{s}{(s-a)(s-b)}$ $(a \neq b)$

   (b) $\dfrac{s}{(s-a)^3}$

   (c) $\dfrac{s}{(s^2+a^2)^2}$

   (d) $\dfrac{s^2+a^2}{(s^2-a^2)^2}$

   (e) $\dfrac{s^3}{(s^2+a^2)^3}$.

2. Show that

$$\mathcal{L}^{-1}\left(\frac{1}{s^2 \cosh s}\right) = t + \frac{8}{\pi^2} \sum_{n=1}^{\infty} \frac{(-1)^n}{(2n-1)^2} \sin\left(\frac{2n-1}{2}\right)\pi t.$$

3. Verify formulas (4.15) and (4.16). [You do not have to verify that

$$\lim_{n \to \infty} \int_{C_n} e^{ts} F(s)\, ds = 0.]$$

4. Show that if $\mathcal{P}_n$ is the parabola given in Example 4.9 and $|F(s)| \leq 1/\left(n + \frac{1}{2}\right)\pi \to 0$ uniformly on $\mathcal{P}_n$ as $n \to \infty$, then

$$\lim_{n \to \infty} \int_{\mathcal{P}_n} e^{ts} F(s)\, ds = 0 \qquad (t > 0).$$

   [Hint: For $x > 0$, i.e., $\tau^2 > \left(n + \frac{1}{2}\right)^2 \pi^2$, show that

$$|ds| = \sqrt{(dx)^2 + (dy)^2} \leq \sqrt{2}\, dx,$$

   and hence,

$$\int_{\mathcal{P}_n(x>0)} |e^{ts}|\, |F(s)|\, |ds| \to 0 \qquad \text{as} \quad n \to \infty.$$

   For $x < 0$, i.e., $\tau^2 < \left(n + \frac{1}{2}\right)^2 \pi^2$, show that

$$|ds| \leq 2\sqrt{2}\left(n + \frac{1}{2}\right)\pi\, d\tau,$$

and hence

$$\int_{P_n(x<0)} |e^{ts}|\,|F(s)|\,|ds| \to 0 \qquad \text{as} \quad n \to \infty.$$

In this case, one also requires the fact that

$$\int_0^m e^{\tau^2 - m^2}\,d\tau \to 0 \qquad \text{as} \quad m \to \infty.]$$

5. Using the complex inversion formula, show that

$$\mathcal{L}^{-1}\left(\frac{1}{s^{\nu}}\right) = \frac{\sin \nu\pi}{\pi}\,\frac{\Gamma(1-\nu)}{t^{1-\nu}}, \qquad \nu > 0$$

Hence by (2.2) deduce the formula

$$\Gamma(\nu)\,\Gamma(1-\nu) = \pi \csc \nu\pi.$$

(Note: For $\nu = 1/2$, this is the branch point example.)

6. Determine $\mathcal{L}\bigl(\mathrm{erfc}(\sqrt{t})\bigr)$.

5
CHAPTER

Partial Differential Equations

Partial differential equations, like their one-variable counterpart, ordinary differential equations, are ubiquitous throughout the scientific spectrum. However, they are, in general, more difficult to solve. Yet here again, we may apply the Laplace transform method to solve PDEs by reducing the initial problem to a simpler ODE.

Partial differential equations come in three types. For a function of two variables $u = u(x, y)$, the general second-order linear PDE has the form

$$a\frac{\partial^2 u}{\partial x^2} + 2b\frac{\partial^2 u}{\partial x\,\partial y} + c\frac{\partial^2 u}{\partial y^2} + d\frac{\partial u}{\partial x} + e\frac{\partial u}{\partial y} + fu = g, \qquad (5.1)$$

where a, b, c, d, e, f, g may depend on x and y only. We call (5.1)

$$
\begin{array}{lll}
\textit{elliptic} & \text{if} & b^2 - ac < 0, \\
\textit{hyperbolic} & \text{if} & b^2 - ac > 0, \\
\textit{parabolic} & \text{if} & b^2 - ac = 0.
\end{array}
$$

Example 5.1.

(i) The *heat equation*

$$\frac{\partial u}{\partial t} = c\frac{\partial^2 u}{\partial x^2}$$

is parabolic.

175

(ii) The *wave equation*

$$\frac{\partial^2 u}{\partial t^2} = a^2 \frac{\partial^2 u}{\partial x^2}$$

is hyperbolic.

(iii) The *Laplace equation*

$$\frac{\partial^2 u}{\partial x^2} + \frac{\partial^2 u}{\partial y^2} = 0$$

is elliptic.

Laplace Transform Method. We consider the function $u = u(x, t)$, where $t \geq 0$ is a time variable. Denote by $U(x, s)$ *the Laplace transform of u with respect to t,* that is to say

$$U(x, s) = \mathcal{L}(u(x, t)) = \int_0^\infty e^{-st} u(x, t)\, dt.$$

Here x is the "untransformed variable."

Example 5.2.

$$\mathcal{L}(e^{a(x+t)}) = \frac{e^{ax}}{s - a}.$$

We will assume that derivatives and limits pass through the transform.

Assumption (1).

$$\mathcal{L}\left(\frac{\partial u}{\partial x}\right) = \int_0^\infty e^{-st} \frac{\partial}{\partial x} u(x, t)\, dt$$

$$= \frac{\partial}{\partial x} \int_0^\infty e^{-st} u(x, t)\, dt$$

$$= \frac{\partial}{\partial x} U(x, s). \tag{5.2}$$

In other words, "the transform of the derivative is the derivative of the transform."

Assumption (2).

$$\lim_{x \to x_0} \int_0^\infty e^{-st} u(x, t)\, dt = \int_0^\infty e^{-st} u(x_0, t)\, dt, \tag{5.3}$$

that is,

$$\lim_{x \to x_0} U(x, s) = U(x_0, s).$$

In (5.2) it is convenient to write

$$\frac{\partial}{\partial x} U(x, s) = \frac{d}{dx} U(x, s) = \frac{dU}{dx},$$

since our parameter s can be treated like a constant with respect to the differentiation involved. A second derivative version of (5.2) results in the expression

$$\mathcal{L}\left(\frac{\partial^2 u}{\partial x^2}\right) = \frac{d^2 U}{dx^2}.$$

Note that in the present context the derivative theorem (2.7) reads

$$\mathcal{L}\left(\frac{\partial u}{\partial t}\right) = s \mathcal{L}(u(x, t)) - u(x, 0^+)$$

$$= s U(x, s) - u(x, 0^+).$$

The Laplace transform method applied to the solution of PDEs consists of first applying the Laplace transform to both sides of the equation as we have done before. This will result in an ODE involving U as a function of the single variable x.

For example, if

$$\frac{\partial u}{\partial x} = \frac{\partial u}{\partial t}, \tag{5.4}$$

then

$$\mathcal{L}\left(\frac{\partial u}{\partial x}\right) = \mathcal{L}\left(\frac{\partial u}{\partial t}\right),$$

implying

$$\frac{d}{dx} U(x, s) = s U(x, s) - u(x, 0^+). \tag{5.5}$$

The ODE obtained is then solved by whatever means avail themselves. If, say, $u(x, 0^+) = x$ for equation (5.4), we find that the general solution is given by

$$U(x, s) = c e^{sx} + \frac{x}{s} + \frac{1}{s^2}. \tag{5.6}$$

PDE problems in physical settings come with one or more boundary conditions, say for (5.4) that

$$u(0, t) = t. \qquad (5.7)$$

Since the boundary conditions also express u as a function of t, we take the rather unusual step of taking the Laplace transform of the boundary conditions as well. So for (5.7)

$$U(0, s) = \mathcal{L}\big(u(0, t)\big) = \frac{1}{s^2}.$$

Feeding this into (5.6) gives $c = 0$ so that

$$U(x, s) = \frac{x}{s} + \frac{1}{s^2}.$$

Since this is the transform of the desired function $u(x, t)$, inverting gives the solution to (5.4) and (5.7) [(and $u(x, 0^+) = x$]:

$$u(x, t) = x + t.$$

This simple example illustrates the basic techniques involved in solving partial differential equations.

In what follows we will demonstrate the utility of the Laplace transform method when applied to a variety of PDEs. However, before proceeding further, we require two more inverses based upon (4.27):

$$\mathcal{L}^{-1}\left(\frac{e^{-a\sqrt{s}}}{s}\right) = \text{erfc}\left(\frac{a}{2\sqrt{t}}\right), \qquad a > 0.$$

Theorem 5.3.

(i) $\mathcal{L}^{-1}(e^{-a\sqrt{s}}) = \dfrac{a}{2\sqrt{\pi\, t^3}}\, e^{-a^2/4t} \qquad (a > 0).$

(ii) $\mathcal{L}^{-1}\left(\dfrac{e^{-a\sqrt{s}}}{\sqrt{s}}\right) = \dfrac{1}{\sqrt{\pi\, t}}\, e^{-a^2/4t} \qquad (a > 0).$

PROOF. (i) Applying the derivative theorem to (4.27) and noting that $\text{erfc}\,(a/2\sqrt{t}) \to 0$ as $t \to 0^+$, we have

$$\mathcal{L}\left(\frac{d}{dt}\, \text{erfc}\left(\frac{a}{2\sqrt{t}}\right)\right) = e^{-a\sqrt{s}},$$

that is,

$$\mathcal{L}\left(\frac{a}{2\sqrt{\pi\, t^3}}\, e^{-a^2/4t}\right) = e^{-a\sqrt{s}}, \tag{5.8}$$

as desired.

For (ii), we differentiate (5.8) with respect to s,

$$\frac{d}{ds}\,\mathcal{L}\left(\frac{a}{2\sqrt{\pi\, t^3}}\, e^{-a^2/4t}\right) = -\frac{a\, e^{-a\sqrt{s}}}{2\sqrt{s}},$$

and by Theorem 1.34,

$$\mathcal{L}\left(-\frac{at}{2\sqrt{\pi\, t^3}}\, e^{-a^2/4t}\right) = -\frac{a\, e^{-a\sqrt{s}}}{2\sqrt{s}},$$

which after cancellation gives (ii). $\qquad\qquad\square$

Example 5.4. Solve the boundary-value problem

$$x\frac{\partial y}{\partial x} + \frac{\partial y}{\partial t} + ay = bx^2, \qquad x > 0,\ t > 0,\ a, b\ \text{constants},\ (5.9)$$

$$y(0, t) = 0, \qquad y(x, 0^+) = 0.$$

Setting $\mathcal{L}\big(y(x, t)\big) = Y(x, s)$ and taking the Laplace transform of both sides of (5.9) give

$$xY_x(x, s) + s\, Y(x, s) - y(x, 0^+) + a\, Y(x, s) = \frac{bx^2}{s},$$

that is,

$$x\frac{dY}{dx} + (s + a)Y = \frac{bx^2}{s},$$

or

$$\frac{dY}{dx} + \frac{(s + a)}{x}\, Y = \frac{bx}{s} \qquad (s > 0).$$

Solving this first-order ODE using an integrating factor gives

$$Y(x, s) = \frac{bx^2}{s(s + a + 2)} + cx^{-(s+a)} \qquad (x > 0, s > -a).$$

Taking the Laplace transform of the boundary condition $y(0, t) = 0$ gives

$$Y(0, s) = \mathcal{L}\big(y(0, t)\big) = 0,$$

and thus $c = 0$. Therefore,

$$Y(x, s) = \frac{bx^2}{s(s + a + 2)},$$

and inverting,

$$y(x, t) = \frac{bx^2}{a + 2}(1 - e^{-(a+2)t})$$

by Example 2.40.

One-Dimensional Heat Equation. The heat flow in a finite or semi-infinite thin rod is governed by the PDE

$$\frac{\partial u}{\partial t} = c\,\frac{\partial^2 u}{\partial x^2},$$

where c is a constant (called the *diffusivity*), and $u(x, t)$ is the temperature at position x and time t. The temperature over a cross-section at x is taken to be uniform. (See Figure 5.1.) Many different scenarios can arise in the solution of the heat equation; we will consider several to illustrate the various techniques involved.

Example 5.5. Solve

$$\frac{\partial^2 u}{\partial x^2} = \frac{\partial u}{\partial t}, \qquad x > 0, \ t > 0, \tag{5.10}$$

for

(i) $u(x, 0^+) = 1, x > 0,$

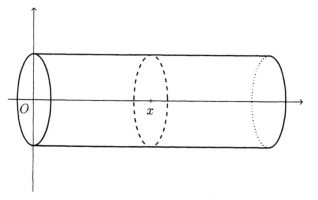

FIGURE 5.1

(ii) $u(0, t) = 0, t > 0,$
(iii) $\lim_{x \to \infty} u(x, t) = 1.$

Taking the Laplace transform of (5.10) yields

$$\frac{d^2 U}{dx^2} = s U - u(x, 0^+) = s U - 1. \tag{5.11}$$

Transforming the boundary conditions (ii) and (iii) gives

$$U(0, s) = \mathcal{L}\big(u(0, t)\big) = 0,$$

$$\lim_{x \to \infty} U(x, s) = \lim_{x \to \infty} \mathcal{L}\big(u(x, t)\big) = \mathcal{L}\big(\lim_{x \to \infty} u(x, t)\big) = \frac{1}{s}.$$

Now (5.11) is an ODE whose solution is given by

$$U(x, s) = c_1 e^{\sqrt{s}x} + c_2 e^{-\sqrt{s}x} + \frac{1}{s}.$$

The boundary condition $\lim_{x \to \infty} U(x, s) = 1/s$ implies $c_1 = 0$, and $U(0, s) = 0$ implies

$$U(x, s) = \frac{1}{s} - \frac{e^{\sqrt{s}x}}{s}.$$

By (4.26),

$$u(x, t) = \operatorname{erf}\left(\frac{x}{2\sqrt{t}}\right) = \frac{2}{\sqrt{\pi}} \int_0^{x/2\sqrt{t}} e^{-u^2} \, du.$$

Direct calculation shows that $u(x, t)$ indeed satisfies (5.10) and that the initial and boundary conditions are satisfied [cf. (2.49)].

Example 5.6. Solve

$$\frac{\partial^2 u}{\partial x^2} = \frac{\partial u}{\partial t}, \qquad x > 0, \ t > 0,$$

for

(i) $u(x, 0^+) = 0,$
(ii) $u(0, t) = f(t), t > 0,$
(iii) $\lim_{x \to \infty} u(x, t) = 0.$

The transformed equation is

$$\frac{d^2 U}{dx^2} - s U = 0,$$

whose solution is given by

$$U(x, s) = c_2 e^{-\sqrt{s}x}$$

in view of condition (iii). By (ii),

$$U(0, s) = \mathcal{L}(f(t)) = F(s),$$

so that $c_2 = F(s)$ and

$$U(x, s) = F(s)e^{-\sqrt{s}x}.$$

Invoking Theorem 5.3 (i) and the convolution theorem 2.39, we have

$$u(x, t) = \int_0^t \frac{x}{2\sqrt{\pi \tau^3}} e^{-x^2/4\tau} f(t - \tau) \, d\tau.$$

Making the substitution $\sigma^2 = x^2/4\tau$, we find that

$$u(x, t) = \frac{2}{\sqrt{\pi}} \int_{x/2\sqrt{t}}^{\infty} e^{-\sigma^2} f\left(t - \frac{x^2}{4\sigma^2}\right) d\sigma,$$

which is the desired solution.

Example 5.7. Solve

$$\frac{\partial^2 u}{\partial x^2} = \frac{\partial u}{\partial t}, \qquad 0 < x < \ell, \ t > 0,$$

for

(i) $u(x, 0^+) = u_0$,

(ii) $\dfrac{\partial}{\partial x} u(0, t) = 0$ (i.e., left end insulated),

(iii) $u(\ell, t) = u_1$.

Taking the Laplace transform gives

$$\frac{d^2 U}{dx^2} = sU - u_0.$$

Then

$$U(x, s) = c_1 \cosh \sqrt{s}x + c_2 \sinh \sqrt{s}x + \frac{u_0}{s},$$

and by (ii), $c_2 = 0$, so that

$$U(x, s) = c_1 \cosh \sqrt{s}x + \frac{u_0}{s}.$$

We find by (iii) that

$$U(\ell, s) = \frac{u_1}{s} = c_1 \cosh \sqrt{s}\,\ell + \frac{u_0}{s},$$

and so

$$c_1 = \frac{u_1 - u_0}{s \cosh \sqrt{s}\,\ell}.$$

Therefore,

$$U(x, s) = \frac{(u_1 - u_0) \cosh \sqrt{s}\,x}{s \cosh \sqrt{s}\,\ell} + \frac{u_0}{s}.$$

Taking the inverse by (4.16) gives

$$u(x, t) = u_0 + (u_1 - u_0)\mathcal{L}^{-1}\left(\frac{\cosh \sqrt{s}\,x}{s \cosh \sqrt{s}\,\ell}\right)$$

$$= u_1 + \frac{4(u_1 - u_0)}{\pi} \sum_{n=1}^{\infty} \frac{(-1)^n}{(2n-1)} e^{-(2n-1)^2 \pi^2 t/4\ell^2}$$

$$\times \cos\left(\frac{2n-1}{2\ell}\right)\pi x.$$

Example 5.8. Solve

$$\frac{\partial^2 u}{\partial x^2} = \frac{\partial u}{\partial t}, \qquad 0 < x < 1, \ t > 0,$$

for

   (i) $u(x, 0^+) = f(x)$,
   (ii) $u(0, t) = 0, \ t > 0$,
  (iii) $u(1, t) = 0, \ t > 0$.

Therefore,

$$\frac{d^2 U}{dx^2} - sU = -f(x).$$

Here we solve this ODE by the Laplace transform method as well. To this end, let $Y(x) = U(x, s)$. Then $Y(0) = U(0, s) = 0$, $Y(1) = U(1, s) = 0$. Setting $a^2 = s$, we obtain

$$\sigma^2 \mathcal{L}(Y) - \sigma Y(0) - Y'(0) - a^2 \mathcal{L}(Y) = -\mathcal{L}(f) = -F(\sigma),$$

that is,

$$\mathcal{L}(Y) = \frac{Y'(0)}{\sigma^2 - a^2} - \frac{F(\sigma)}{\sigma^2 - a^2}.$$

Inverting gives

$$Y(x) = U(x, s) = \frac{Y'(0)\sinh ax}{a} - \frac{1}{a}\int_0^x f(u)\sinh a(x - u)\,du$$

$$= \frac{Y'(0)\sinh\sqrt{s}x}{\sqrt{s}} - \frac{1}{\sqrt{s}}\int_0^x f(u)\sinh\sqrt{s}(x - u)\,du.$$

Now, $Y(1) = 0$, implying

$$Y'(0) = \frac{1}{\sinh\sqrt{s}}\int_0^1 f(u)\sinh\sqrt{s}(1 - u)\,du.$$

Thus,

$$U(x, s) = \int_0^1 f(u)\frac{\sinh\sqrt{s}x\sinh\sqrt{s}(1 - u)}{\sqrt{s}\sinh\sqrt{s}}\,du$$

$$- \int_0^x f(u)\frac{\sinh\sqrt{s}(x - u)}{\sqrt{s}}\,du.$$

We can write $\int_0^1 = \int_0^x + \int_x^1$ and use the fact from Section 3.2 that

$$\sinh(z \pm w) = \sinh z \cosh w \pm \cosh z \sinh w.$$

Then

$$U(x, s) = \int_0^x f(u)\left[\frac{\sinh\sqrt{s}x\sinh\sqrt{s}(1 - u)}{\sqrt{s}\sinh\sqrt{s}} - \frac{\sinh\sqrt{s}(x - u)}{\sqrt{s}}\right]du$$

$$+ \int_x^1 f(u)\frac{\sinh\sqrt{s}x\sinh\sqrt{s}(1 - u)}{\sqrt{s}\sinh\sqrt{s}}\,du$$

$$= \int_0^x f(u)\frac{\sinh\sqrt{s}(1 - x)\sinh\sqrt{s}u}{\sqrt{s}\sinh\sqrt{s}}\,du$$

$$+ \int_x^1 f(u)\frac{\sinh\sqrt{s}x\sinh\sqrt{s}(1 - u)}{\sqrt{s}\sinh\sqrt{s}}\,du.$$

To determine the inverse we use the complex inversion formula. When it is applied to the first integral we have

$$\frac{1}{2\pi i} \int_{x_0-i\infty}^{x_0+i\infty} e^{ts} \left\{ \int_0^x f(u) \frac{\sinh\sqrt{s}(1-x)\sinh\sqrt{s}\,u}{\sqrt{s}\sinh\sqrt{s}}\,du \right\} ds = \Sigma \text{ Res.}$$

There are simple poles in this case at $s_0 = 0$ and $s_n = -n^2\pi^2$, $n = 1, 2, 3, \ldots$ (see Example 4.9).

$$\text{Res}(0) = \lim_{s\to 0} s \int_0^x f(u) \frac{\sinh\sqrt{s}(1-x)\sinh\sqrt{s}\,u}{\sqrt{s}\sinh\sqrt{s}}\,du = 0.$$

$\text{Res}(-n^2\pi^2)$

$$= \lim_{s\to -n^2\pi^2} (s+n^2\pi^2)e^{ts} \int_0^x f(u) \frac{\sinh\sqrt{s}(1-x)\sinh\sqrt{s}\,u}{\sqrt{s}\sinh\sqrt{s}}\,du$$

$$= \lim_{s\to -n^2\pi^2} \frac{s+n^2\pi^2}{\sinh\sqrt{s}} \cdot \lim_{s\to -n^2\pi^2} e^{ts} \int_0^x f(u) \frac{\sinh\sqrt{s}(1-x)\sinh\sqrt{s}\,u}{\sqrt{s}}\,du$$

$$= 2e^{-n^2\pi^2 t} \int_0^x f(u) \frac{\sinh[(n\pi i)(1-x)]\sinh(n\pi i)u}{\cosh(n\pi i)}\,du$$

$$= 2e^{-n^2\pi^2 t} \int_0^x f(u) \frac{\sin[n\pi(1-x)]\sin n\pi u}{-\cos n\pi}\,du,$$

where we have used the properties from Section 3.2 (for $z = x + iy$)

$$\sinh z = \cos y \sinh x + i \sin y \cosh x,$$

$$\cosh z = \cos y \cosh x + i \sin y \sinh x$$

to obtain the last equality.

Therefore,

$$\sum \text{Res} = 2 \sum_{n=1}^{\infty} e^{-n^2\pi^2 t} \left(\int_0^x f(u) \sin n\pi u\,du \right) \sin n\pi x.$$

Similarly, the inverse of the second integral is given by

$$2 \sum_{n=1}^{\infty} e^{-n^2\pi^2 t} \left(\int_x^1 f(u) \sin n\pi u\,du \right) \sin n\pi x.$$

Finally,

$$u(x, t) = 2 \sum_{n=1}^{\infty} e^{-n^2\pi^2 t} \left(\int_0^1 f(u) \sin n\pi u \, du \right) \sin n\pi x.$$

The same result is obtained when we solve this problem by the separation-of-variables method.

One-Dimensional Wave Equation. The wave motion of a string initially lying on the x-axis with one end at the origin can be described by the equation

$$\frac{\partial^2 y}{\partial t^2} = a^2 \frac{\partial^2 y}{\partial x^2}, \qquad x > 0, \ t > 0$$

(Figure 5.2). The displacement is only in the vertical direction and is given by $y(x, t)$ at position x and time t. The constant a is given by $a = \sqrt{T/\rho}$, where T is the tension on the string and ρ its mass per unit length. The same equation happens to describe the longitudinal vibrations in a horizontal beam, where $y(x, t)$ represents the longitudinal displacement of a cross section at x and time t.

Example 5.9. Solve

$$\frac{\partial^2 y}{\partial t^2} = a^2 \frac{\partial^2 y}{\partial x^2}, \qquad x > 0, \ t > 0,$$

for

(i) $y(x, 0^+) = 0, \ x > 0,$

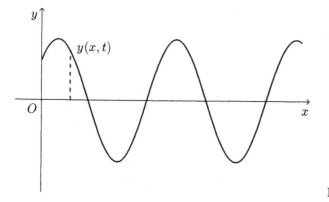

FIGURE 5.2

(ii) $y_t(x, 0^+) = 0$, $x > 0$,
(iii) $y(0, t) = f(t)$ $(f(0) = 0)$,
(iv) $\lim\limits_{x \to \infty} y(x, t) = 0$.

The transformed equation becomes

$$s^2 Y(x, s) - s\, y(x, 0^+) - \frac{\partial}{\partial t}\, y(x, 0^+) = a^2 \frac{d^2 Y}{dx^2},$$

that is,

$$\frac{d^2 Y}{dx^2} - \frac{s^2}{a^2} Y = 0.$$

Solving,

$$Y(x, s) = c_1 e^{(s/a)x} + c_2 e^{-(s/a)x}.$$

Since $y(x, t) \to 0$ as $x \to \infty$, then $c_1 = 0$ and

$$Y(x, s) = c_2 e^{-(s/a)x}.$$

By condition (iii), $Y(0, s) = \mathcal{L}\big(f(t)\big) = F(s)$, so that $c_2 = F(s)$, and

$$Y(x, s) = F(s)\, e^{-(s/a)x}.$$

Inverting via the second translation theorem (1.31) gives

$$y(x, t) = u_{\frac{x}{a}}(t) f\left(t - \frac{x}{a}\right),$$

or

$$y(x, t) = \begin{cases} f\left(t - \frac{x}{a}\right) & t \geq \frac{x}{a} \\ 0 & t < \frac{x}{a}. \end{cases}$$

Thus, the string remains at rest until the time $t = x/a$, after which it exhibits the same motion as the end at $x = 0$, with a time delay of x/a.

Example 5.10. Solve

$$\frac{\partial^2 y}{\partial t^2} = \frac{\partial^2 y}{\partial x^2}, \qquad 0 < x < \ell, \; t > 0,$$

for

(i) $y(0, t) = 0$, $t > 0$,
(ii) $y(\ell, t) = a$, $t > 0$,

(iii) $y(x, 0^+) = 0, 0 < x < \ell$,
(iv) $y_t(x, 0^+) = 0, 0 < x < \ell$.

By transforming the equation we obtain

$$\frac{d^2 Y}{dx^2} - s^2 Y = 0,$$

whose solution is given by

$$Y(x, s) = c_1 \cosh sx + c_2 \sinh sx.$$

Then $0 = Y(0, s) = c_1$ and $Y(x, s) = c_2 \sinh sx$. Moreover,

$$Y(\ell, s) = \frac{a}{s} = c_2 \sinh s\ell$$

and $c_2 = a/s \sinh s\ell$. Thus,

$$Y(x, s) = \frac{a \sinh sx}{s \sinh s\ell}.$$

This function has simple poles at $s_n = n\pi i/\ell$, $n = 0, \pm 1, \pm 2, \ldots$.

$$\text{Res}(0) = \lim_{s \to 0} s e^{ts} \frac{a \sinh sx}{s \sinh s\ell}$$

$$= a \lim_{s \to 0} \frac{x \cosh sx}{\ell \cosh s\ell}$$

$$= \frac{ax}{\ell}.$$

For $n = \pm 1, \pm 2, \cdots$,

$$\text{Res}\left(\frac{n\pi i}{\ell}\right) = a \lim_{s \to \frac{n\pi i}{t}} \left(s - \frac{n\pi i}{\ell}\right) \frac{e^{ts} \sinh sx}{s \sinh s\ell}$$

$$= a \lim_{s \to \frac{n\pi i}{t}} \frac{\left(s - \frac{n\pi i}{\ell}\right)}{\sinh s\ell} \cdot \lim_{s \to \frac{n\pi i}{t}} \frac{e^{ts} \sinh sx}{s}$$

$$= \frac{a}{\ell \cosh n\pi i} \frac{e^{n\pi i t/\ell} \sinh \frac{n\pi ix}{\ell}}{n\pi i/\ell}$$

$$= \frac{a}{n\pi}(-1)^n e^{n\pi i t/\ell} \sin \frac{n\pi x}{\ell}.$$

Therefore,

$$y(x, t) = \sum \operatorname{Res} = \frac{ax}{\ell} + \sum_{\substack{n=-\infty \\ n \neq 0}}^{\infty} (-1)^n \frac{a}{n\pi} e^{n\pi i t/\ell} \sin \frac{n\pi x}{\ell}$$

$$= \frac{ax}{\ell} + \frac{2a}{\pi} \sum_{n=1}^{\infty} \frac{(-1)^n}{n} \sin \frac{n\pi x}{\ell} \cos \frac{n\pi t}{\ell},$$

by the complex inversion formula.

Exercises

1. Solve the boundary-value problem

$$\frac{\partial y}{\partial x} + \frac{1}{x} \frac{\partial y}{\partial t} = t, \qquad x > 0, \ t > 0,$$

$$y(x, 0^+) = x, \qquad y(0, t) = 0.$$

2. Solve the following heat equations.

(a) $\dfrac{\partial^2 u}{\partial x^2} = \dfrac{\partial u}{\partial t}$, $\qquad x > 0, \ t > 0,$

$\quad$ **(i)** $u(x, 0^+) = 0, \quad x > 0,$
$\quad$ **(ii)** $u(0, t) = \delta(t), \quad t > 0,$
$\quad$ **(iii)** $\lim\limits_{x \to \infty} u(x, t) = 0.$

(b) $\dfrac{\partial^2 u}{\partial x^2} = \dfrac{\partial u}{\partial t}$, $\qquad x > 0, \ t > 0,$

$\quad$ **(i)** $u(x, 0^+) = u_0, \quad x > 0,$
$\quad$ **(ii)** $u(0, t) = u_1, \quad t > 0,$
$\quad$ **(iii)** $\lim\limits_{x \to \infty} u(x, t) = u_0.$

(c) $\dfrac{\partial^2 u}{\partial x^2} = \dfrac{\partial u}{\partial t}$, $\qquad 0 < x < 1, \ t > 0,$

$\quad$ **(i)** $u(x, 0^+) = 0, \quad 0 < x < 1,$
$\quad$ **(ii)** $u(0, t) = 0, \quad t > 0,$
$\quad$ **(iii)** $u(1, t) = 1, \quad t > 0.$

(d) $\dfrac{\partial^2 u}{\partial x^2} = \dfrac{\partial u}{\partial t}$, $0 < x < \ell, \ t > 0$,

   (i) $u(x, 0^+) = ax$, $0 < x < \ell$ (a constant),
   (ii) $u(0, t) = 0$, $t > 0$,
   (iii) $u(\ell, t) = 0$, $t > 0$.

3. Solve the following wave equations.

(a) $\dfrac{\partial^2 y}{\partial t^2} = \dfrac{\partial^2 y}{\partial x^2}$, $0 < x < 1, \ t > 0$,

   (i) $y(x, 0^+) = \sin \pi x$, $0 < x < 1$,
   (ii) $y(0, t) = 0$, $t > 0$,
   (iii) $y(1, t) = 0$, $t > 0$,
   (iv) $y_t(x, 0) = 0$, $0 < x < 1$.

(b) $\dfrac{\partial^2 y}{\partial t^2} = \dfrac{\partial^2 y}{\partial x^2}$, $0 < x < 1, \ t > 0$,

   (i) $y(x, 0^+) = 0$, $0 < x < 1$,
   (ii) $y(1, t) = 1$, $t > 0$,
   (iii) $y_x(0, t) = 0$, $t > 0$,
   (iv) $y_t(x, 0^+) = 0$, $0 < x < 1$.

(c) $\dfrac{\partial^2 y}{\partial t^2} = \dfrac{\partial^2 y}{\partial x^2}$, $0 < x < 1, \ t > 0$,

   (i) $y(x, 0^+) = 0$, $0 < x < 1$,
   (ii) $y(0, t) = 0$, $t > 0$,
   (iii) $y(1, t) = 0$, $t > 0$,
   (iv) $y_t(x, 0^+) = x$, $0 < x < 1$.

(d) $\dfrac{\partial^2 y}{\partial t^2} = \dfrac{\partial^2 y}{\partial x^2}$, $0 < x < 1, \ t > 0$, for

   (i) $y(x, 0^+) = f(x)$, $x > 0$,
   (ii) $y(0, t) = 0$, $t > 0$,
   (iii) $y(1, t) = 0$, $t > 0$,
   (iv) $y_t(x, 0^+) = 0$, $0 < x < 1$.

(Note: This problem is similar to Example 5.8.)

4. Solve the boundary-value problem

$$\frac{\partial^2 y}{\partial t^2} = \frac{\partial^2 y}{\partial x^2} - \sin \pi x, \qquad 0 < x < 1, \ t > 0,$$

for

(i) $y(x, 0^+) = 0, \quad 0 < x < 1,$
(ii) $y(0, t) = 0, \quad t > 0,$
(iii) $y(1, t) = 0, \quad t > 0,$
(iv) $y_t(x, 0^+) = 0, \quad 0 < x < 1.$

5. A "fundamental solution" to the heat equation satisfies

$$\frac{\partial u}{\partial t} = a^2 \frac{\partial^2 u}{\partial x^2}, \qquad x > 0, \ t > 0, \ a > 0,$$

for

(i) $u(x, 0^+) = \delta(x), \ x > 0,$

(iv) $\dfrac{\partial}{\partial x} u(0, t) = 0, \quad t > 0,$

(iii) $\lim\limits_{x \to \infty} u(x, t) = 0.$

Solve for $u(x, t)$. (See Exercises 2.5, Question 7.)

Appendix

The sole integral used in this text is the Riemann integral defined as follows.

Let

$$\Delta = \{a = t_0 < t_1 < \cdots < t_n = b\}$$

be a *partition* of the interval $[a, b]$. Let f be a function defined on $[a, b]$ and choose any point $x_i \in [t_{i-1}, t_i]$, $i = 1, \cdots, n$. The sum

$$\sum_{i=1}^{n} f(x_i)(t_i - t_{i-1})$$

is called a *Riemann sum*. Denote by $\|\Delta\| = \max_{1 \le i \le n}(t_i - t_{i-1})$.

The function f is said to be *Riemann integrable* if there is a number I_{ab} such that for any $\varepsilon > 0$, there exists a $\delta > 0$ such that for each partition Δ of $[a, b]$ with $\|\Delta\| < \delta$, we have

$$\left| \sum_{i=1}^{n} f(x_i)(t_i - t_{i-1}) - I_{ab} \right| < \varepsilon,$$

for all choices of $x_i \in [t_{i-1}, t_i]$, $i = 1, \cdots, n$. The value I_{ab} is the *Riemann integral of f over* $[a, b]$ and is written as

$$I_{ab} = \int_a^b f(t)\, dt.$$

It is worth noting that *if f is Riemann integrable on* $[a, b]$, *it is bounded on* $[a, b]$. Moreover, *every continuous function on* $[a, b]$ *is Riemann integrable there.*

In order to see just how dangerous it can be to pass a limit inside an integral without sound justification, consider the following.

Example A.1. Let $\{f_n\}$ be a sequence of functions defined on $[0, 1]$ by

$$f_n(t) = \begin{cases} 4n^2 t & 0 \le t \le \dfrac{1}{2n} \\[2ex] -4n^2 t + 4n & \dfrac{1}{2n} < t < \dfrac{1}{n} \\[2ex] 0 & \dfrac{1}{n} \le t \le 1 \end{cases}$$

(Figure A.1). Since $f_n(0) = 0$, $\lim_{n \to 0} f_n(0) = 0$. Moreover, for $t > 0$ and $n > 1/t$, $f_n(t) = 0$, implying

$$\lim_{n \to 0} f_n(t) = 0, \qquad t \in [0, 1].$$

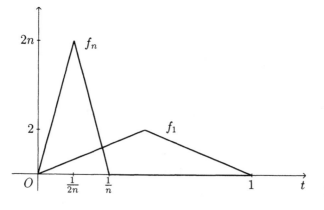

FIGURE A.1

By construction,

$$\int_0^1 f_n(t)\, dt = 1,$$

so that $\lim_{n\to\infty} \int_0^1 f_n(t)\, dt = 1$. On the other hand,

$$\int_0^1 \lim_{n\to\infty} f_n(t)\, dt = \int_0^1 0\, dt = 0.$$

In Theorem 3.1 it was shown that the Laplace transform of a piecewise continuous function of exponential order is an analytic function. A necessary ingredient in that proof was that the Laplace transform be continuous.

Theorem A.2. *If f is a piecewise continuous function, and*

$$\int_0^\infty e^{-st} f(t)\, dt = F(s)$$

converges uniformly for all $s \in E \subseteq \mathbb{C}$, then $F(s)$ is a continuous function on E, that is, for $s \to s_0 \in E$,

$$\lim_{s\to s_0} \int_0^\infty e^{-st} f(t)\, dt = \int_0^\infty \lim_{s\to s_0} e^{-st} f(t)\, dt = F(s_0).$$

PROOF. By the uniform convergence of the integral, given $\varepsilon > 0$ there exists some t_0 such that for all $\tau \geq t_0$,

$$\left| \int_\tau^\infty e^{-st} f(t)\, dt \right| < \varepsilon, \tag{A.1}$$

for all $s \in E$.

Now consider

$$\left| \int_0^\infty e^{-st} f(t)\, dt - \int_0^\infty e^{-s_0 t} f(t)\, dt \right| = \left| \int_0^\infty (e^{-st} - e^{-s_0 t}) f(t)\, dt \right|$$

$$\leq \int_0^{t_0} |e^{-st} - e^{-s_0 t}|\, |f(t)|\, dt + \left| \int_{t_0}^\infty (e^{-st} - e^{-s_0 t}) f(t)\, dt \right|.$$

In view of (A.1), the second integral satisfies

$$\left| \int_{t_0}^\infty (e^{-st} - e^{-s_0 t}) f(t)\, dt \right| \leq \left| \int_{t_0}^\infty e^{-st} f(t)\, dt \right| + \left| \int_{t_0}^\infty e^{-s_0 t} f(t)\, dt \right|$$

$$< \varepsilon + \varepsilon = 2\varepsilon.$$

For the first integral,

$$\int_0^{t_0} |e^{-st} - e^{-s_0 t}| \, |f(t)| \, dt \le M \int_0^{t_0} |e^{-st} - e^{-s_0 t}| \, dt$$

since f is piecewise continuous, hence bounded on $[0, t_0]$. Finally, $|e^{-st} - e^{-s_0 t}|$ can be made uniformly small for $0 \le t \le t_0$ and s sufficiently close to s_0,* say

$$|e^{-st} - e^{-s_0 t}| < \frac{1}{M t_0} \varepsilon.$$

Hence

$$\int_0^{t_0} |e^{-st} - e^{-s_0 t}| \, |f(t)| \, dt < \varepsilon,$$

and so

$$\lim_{s \to s_0} \int_0^\infty e^{-st} f(t) \, dt = \int_0^\infty e^{-s_0 t} f(t) \, dt. \qquad \square$$

A more subtle version of the preceding result which was used in the proof of the terminal-value theorem (2.36) is the following.

Theorem A.3. *Suppose that f is piecewise continuous on $[0, \infty)$ and $\mathcal{L}(f(t)) = F(s)$ exists for all $s > 0$, and $\int_0^\infty f(t) \, dt$ converges. Then*

$$\lim_{s \to 0^+} \int_0^\infty e^{-st} f(t) \, dt = \int_0^\infty f(t) \, dt.$$

PROOF. Since $\int_0^\infty f(t) \, dt$ converges, given $\varepsilon > 0$, fix τ_0 sufficiently large so that

$$\left| \int_{\tau_0}^\infty f(t) \, dt \right| < \frac{\varepsilon}{2}. \qquad (A.2)$$

Next consider

$$\left| \int_0^\infty f(t) \, dt - \int_0^\infty e^{-st} f(t) \, dt \right| = \left| \int_0^\infty (1 - e^{-st}) f(t) \, dt \right|$$

$$\le \int_0^{\tau_0} (1 - e^{-st}) |f(t)| \, dt + \int_{\tau_0}^\infty (1 - e^{-st}) |f(t)| \, dt.$$

---

*We are using the fact that the function $g(s, t) = e^{-st}$ is *uniformly continuous* on a closed rectangle.

For the first integral, since f is piecewise continuous it is bounded on $[0, \tau_0]$, say $|f(t)| \leq M$. Then

$$\int_0^{\tau_0} (1 - e^{-st})|f(t)| \, dt \leq M \int_0^{\tau_0} (1 - e^{-st}) \, dt$$

$$= \frac{M(s\tau_0 + e^{-s\tau_0} - 1)}{s} \to 0$$

as $s \to 0^+$ by an application of l'Hôpital's rule. Thus the first integral can be made smaller than $\varepsilon/2$ for s sufficiently small.

For the second integral

$$\int_{\tau_0}^\infty (1 - e^{-st})|f(t)| \, dt \leq \int_{\tau_0}^\infty |f(t)| \, dt < \frac{\varepsilon}{2}$$

by $(A.2)$. Therefore,

$$\left| \int_0^\infty f(t) \, dt - \int_0^\infty e^{-st} f(t) \, dt \right| < \varepsilon$$

for all s sufficiently small, proving the result. $\qquad\square$

Corollary A.4. *Suppose that f satisfies the conditions of the derivative theorem (2.7), $\mathcal{L}(f'(t)) = F(s)$ exists for all $s > 0$, and $\lim_{t\to\infty} f(t)$ exists. Then*

$$\lim_{s\to 0^+} \int_0^\infty e^{-st} f'(t) \, dt = \int_0^\infty f'(t) \, dt.$$

PROOF. Note that f' is piecewise continuous on $[0, \infty)$ and as we have shown in the proof of Theorem 2.36 [namely, equation (2.47)], the existence of $\lim_{t\to\infty} f(t)$ implies that $\int_0^\infty f'(t) \, dt$ converges. The result now follows from the theorem. $\qquad\square$

Even though a function is only piecewise continuous, its integral is continuous.

Theorem A.5. *If f is piecewise continuous on $[0, \infty)$, then the function*

$$g(t) = \int_0^t f(u) \, du$$

is continuous on $[0, \infty)$.

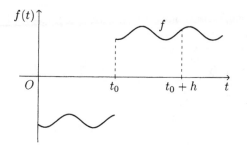

FIGURE A.2

PROOF. Assume that t_0 is a point of discontinuity of f (Figure A.2). Then for $h > 0$,

$$g(t_0 + h) - g(t_0) = \int_{t_0}^{t_0+h} f(u)\, du. \qquad (A.3)$$

Since f is piecewise continuous, f is bounded on $(t_0, t_0 + h)$, say $|f| < M$ there. Thus we find that

$$\left| \int_{t_0}^{t_0+h} f(u)\, du \right| < M \int_{t_0}^{t_0+h} du = Mh \to 0 \qquad (A.4)$$

as $h \to 0$. In view of (A.3), we obtain

$$\lim_{h \to 0^+} g(t_0 + h) = g(t_0).$$

Similarly,

$$\lim_{h \to 0^-} g(t_0 + h) = g(t_0)$$

for $t_0 > 0$.

If t_0 is a point of continuity of f, the proof is the same. □

The justification of applying the Laplace transform method to solving ODEs is aided by the fact that the solution will be continuous of exponential order and thus possess a Laplace transform.

Theorem A.6. *For the nth-order, linear, nonhomogeneous ordinary differential equation*

$$a_n y^{(n)} + a_{n-1} y^{(n-1)} + \cdots + a_0 y = f(t), \qquad (A.5)$$

$a_0, a_1, \ldots, a_n$ *constants, if $f(t)$ is continuous on $[0, \infty)$ and of exponential order, then any solution is also continuous on $[0, \infty)$ of exponential order.*

PROOF. We give the proof for $n = 2$ as the proof for higher-order equations follows similarly.

For

$$ay'' + by' + cy = f(t),$$

the general solution y_h to the corresponding homogeneous equation $ay'' + by' + cy = 0$ is given by

$$y_h = c_1 y_1 + c_2 y_2,$$

where y_1, y_2 are two linearly independent solutions (of the homogeneous equation) of the prescribed form (cf., e.g., Zill [16])

(i) $e^{m_1 t}$, $e^{m_2 t}$,
(ii) e^{mt}, $t\, e^{mt}$,
(iii) $e^{at} \cos bt$, $e^{at} \sin bt$.

Since each of these terms has exponential order, y_h does also and is continuous on $[0, \infty)$.

A particular solution y_p of (A.5) can be found by the *method of variation of parameters* (Zill [16]). Here y_p takes the form

$$y_p = u_1 y_1 + u_2 y_2,$$

where

$$u_1' = \frac{-y_2 f(t)}{a W(y_1, y_2)}, \qquad u_2' = \frac{y_1 f(t)}{a W(y_1, y_2)},$$

and $W(y_1, y_2)$ is the *Wronskian*

$$W(y_1, y_2) = \begin{vmatrix} y_1 & y_2 \\ y_1' & y_2' \end{vmatrix} = y_1 y_2' - y_1' y_2 \neq 0.$$

In cases (i), (ii), (iii), $W(y_1, y_2)$ can be determined and seen to be of the form $M e^{\beta t}$ and hence so is $W^{-1}(y_1, y_2)$. Since the product of functions of exponential order also have exponential order [Exercises 1.4, Question 1(ii)], we conclude that u_1' and u_2' have exponential order and are continuous on $[0, \infty)$. The same holds for u_1 and u_2 by Remark 2.11 and likewise for y_p.

Finally, the general solution of (A.5), given by

$$y = y_h + y_p,$$

is continuous and has exponential order, as desired. $\square$

Remark. Let us show that for $n = 2$, the Laplace transform method is indeed justified in solving (A.5) under the conditions stipulated. In fact, not only is $y = y_h + y_p$ continuous and of exponential order on $[0, \infty)$, but so is

$$y' = y_h' + y_p' = y_h' + (u_1'y_1 + u_1y_1' + u_2'y_2 + u_2y_2'),$$

and hence also $y'' = (1/a)(f(t) - by' - cy)$. The hypotheses of Theorem 2.12 are clearly satisfied, and the Laplace transform method can be applied.

In general, the continuity of

$$y(t), y'(t), \ldots, y^{(n-1)}(t)$$

for $t > 0$ is a basic a priori requirement of a unique solution to (A.5) (see Doetsch [3], p. 78).

A useful result in dealing with partial fraction decompositions is the following

Theorem A.7 (Fundamental Theorem of Algebra). *Every polynomial of degree n,*

$$p(z) = a_nz^n + a_{n-1}z^{n-1} + \cdots + a_1z + a_0, \qquad a_n \neq 0,$$

with complex coefficients, has exactly n roots in $\mathbb{C}$, with the roots counted according to multiplicity.

Corollary A.8. *Any two polynomials of degree n that are equal at $n + 1$ points are identically equal.*

This is so because the difference of the two polynomials is itself a polynomial of degree n and therefore can vanish at n points only unless it is identically the zero polynomial, in which case all the coefficients must be zero. Thus the two polynomials have identical coefficients.

The Riemann–Stieltjes integral was introduced in Section 2.5 in order to deal with the Laplace transform of the Dirac distribution. It enjoys properties similar to those of the conventional Riemann integral.

Theorem A.9.

(i) If $\int_a^b f d\varphi_1$ and $\int_a^b f d\varphi_2$ both exist, and $\varphi = \varphi_1 + \varphi_2$, then f is (Riemann–Stieltjes) integrable with respect to φ and

$$\int_a^b f d\varphi = \int_a^b f d\varphi_1 + \int_a^b f d\varphi_2.$$

(ii) If $\int_a^b f_1 d\varphi$ and $\int_a^b f_2 d\varphi$ both exist and $f = f_1 + f_2$, then f is integrable with respect to φ and

$$\int_a^b f d\varphi = \int_a^b f_1 d\varphi + \int_a^b f_2 d\varphi.$$

(iii) If $\int_a^b f d\varphi$ exists, then for any constant c,

$$\int_a^b (cf) d\varphi = c \int_a^b f d\varphi.$$

(iv) If $\int_a^c f d\varphi$ and $\int_c^b f d\varphi$ exist, $a < c < b$, then $\int_a^b f d\varphi$ exists and

$$\int_a^b f d\varphi = \int_a^c f d\varphi + \int_c^b f d\varphi.$$

The proofs are a natural consequence of the definition of the Riemann–Stieltjes integral.

Theorem A.10. If f, φ, φ' are continuous on $[a, b]$, then $\int_a^b f d\varphi$ exists and

$$\int_a^b f(t) \, d\varphi(t) = \int_a^b f(t) \varphi'(t) \, dt.$$

PROOF. Given $\varepsilon > 0$, we need to show that

$$\left| \sum_{j=1}^n f(x_j)[\varphi(t_j) - \varphi(t_{j-1})] - \int_a^b f(t) \varphi'(t) \, dt \right| < \varepsilon \qquad \text{(A.6)}$$

for $\Delta = \max_j (t_j - t_{j-1})$ sufficiently small. By the mean-value theorem, we can express the left-hand side as

$$\sum_{j=1}^n f(x_j)[\varphi(t_j) - \varphi(t_{j-1})] = \sum_{j=1}^n f(x_j) \varphi'(\xi_j)(t_j - t_{j-1}) \qquad \text{(A.7)}$$

for some $\xi_j \in [t_{j-1}, t_j]$. The right-hand side is nearly what we require.

Since f is continuous on $[a, b]$, let $|f(t)| \leq M$, $t \in [a, b]$. Now φ' is continuous on $[a, b]$ and hence uniformly continuous there. Consequently, there exists some $\delta > 0$ such that whenever $|\xi_j - x_j| < \delta$, it follows that

$$|\varphi'(\xi_j) - \varphi'(x_j)| < \frac{\varepsilon}{2M(b-a)}. \tag{A.8}$$

Since $f\varphi'$ is Riemann integrable, for any suitably fine subdivision of $[a, b]$, with $\Delta < \delta$, we have

$$\left| \sum_{j=1}^{n} f(x_j) \varphi'(x_j)(t_j - t_{j-1}) - \int_a^b f(t) \varphi'(t)\, dt \right| < \frac{\varepsilon}{2}. \tag{A.9}$$

From (A.8) we get

$$\left| \sum_{j=1}^{n} f(x_j)[\varphi'(\xi_j) - \varphi'(x_j)](t_j - t_{j-1}) \right|$$

$$< \sum_{j=1}^{n} M \left| \frac{\varepsilon}{2M(b-a)} (t_j - t_{j-1}) \right| = \frac{\varepsilon}{2}. \tag{A.10}$$

Finally, taking (A.9) and (A.10) together, with ξ_j, $x_j \in [t_{j-1}, t_j]$, and with the triangle inequality, gives

$$\left| \sum_{j=1}^{n} f(x_j) \varphi'(\xi_j)(t_j - t_{j-1}) - \int_a^b f(t) \varphi'(t)\, dt \right| < \varepsilon.$$

In view of (A.7), we have established (A.6). □

In order to reverse the order of integration, as in Theorem 1.37, we use the next result.

Theorem A.11. *If $f(x, t)$ is continuous on each rectangle $a \leq x \leq b$, $0 \leq t \leq T$, $T > 0$, except for possibly a finite number of jump discontinuities across the lines $t = t_i$, $i = 1, \ldots, n$, and if $\int_0^\infty f(x, t)\, dt$ converges uniformly for all x in $[a, b]$, then*

$$\int_a^b \int_0^\infty f(x, t)\, dt\, dx = \int_0^\infty \int_a^b f(x, t)\, dx\, dt.$$

PROOF. From the theory of ordinary integrals we have

$$\int_0^\tau \int_a^b f(x, t)\,dx\,dt = \int_a^b \int_0^\tau f(x, t)\,dt\,dx,$$

implying

$$\int_0^\infty \int_a^b f(x, t)\,dx\,dt = \lim_{\tau \to \infty} \int_a^b \int_0^\tau f(x, t)\,dt\,dx. \qquad (A.11)$$

For the other integral

$$\int_a^b \int_0^\infty f(x, t)\,dt\,dx = \int_a^b \int_0^\tau f(x, t)\,dt\,dx + \int_a^b \int_\tau^\infty f(x, t)\,dt\,dx. \qquad (A.12)$$

Since $\int_0^\infty f(x, t)\,dt$ converges uniformly, given any $\varepsilon > 0$, there exists $T > 0$ such that for all $\tau \geq T$

$$\left| \int_\tau^\infty f(x, t)\,dt \right| < \frac{\varepsilon}{b - a},$$

for all x in $[a, b]$. Hence for $\tau \geq T$,

$$\left| \int_a^b \int_\tau^\infty f(x, t)\,dt\,dx \right| < \varepsilon,$$

that is,

$$\lim_{\tau \to \infty} \int_a^b \int_\tau^\infty f(x, t)\,dt\,dx = 0.$$

Letting $\tau \to \infty$ in (A.12),

$$\int_a^b \int_0^\infty f(x, t)\,dt\,dx = \int_0^\infty \int_a^b f(x, t)\,dx\,dt$$

via (A.11). □

Note that the hypotheses are satisfied by our typical integrand $e^{-xt}f(t)$, where f is piecewise continuous of exponential order.

The following general theorem tells when taking the derivative inside an integral such as the Laplace integral, is justified.

Theorem A.12. *Suppose that $f(x, t)$ and $\partial/\partial x f(x, t)$ are continuous on each rectangle $a \leq x \leq b, 0 \leq t \leq T, T > 0$, except possibly for a* *finite number of jump discontinuities across the lines* $t = t_i$, $i = 1, \ldots, n$, *and of the two integrals*

$$F(x) = \int_0^\infty f(x, t)\, dt \quad \text{and} \quad \int_0^\infty \frac{\partial}{\partial x} f(x, t)\, dt;$$

the first one converges and the second one converges uniformly. Then

$$\frac{d}{dx} F(x) = \int_0^\infty \frac{\partial}{\partial x} f(x, t)\, dt \qquad (a < x < b).$$

PROOF. Let

$$G(u) = \int_0^\infty \frac{\partial}{\partial u} f(u, t)\, dt.$$

Then G is continuous as in the proof of Theorem A.2 and employing Theorem A.11 gives

$$\int_a^x G(u)\, du = \int_a^x \int_0^\infty \frac{\partial}{\partial u} f(u, t)\, dt\, du$$

$$= \int_0^\infty \int_a^x \frac{\partial}{\partial u} f(u, t)\, du\, dt$$

$$= \int_0^\infty [f(x, t) - f(a, t)]\, dt$$

$$= F(x) - F(a).$$

Therefore,

$$\frac{d}{dx} F(x) = G(x) = \int_0^\infty \frac{\partial}{\partial x} f(x, t)\, dt. \qquad \square$$

A consequence of the preceding theorem which was useful in Chapter 4 follows.

Theorem A.13.

$$\frac{1}{\pi} \int_0^\infty \frac{e^{-tx} \sin a\sqrt{x}}{x}\, dx = \mathrm{erf}\left(\frac{a}{2\sqrt{t}}\right), \qquad t > 0.$$

PROOF. Denote the left-hand side by $y(a, t)$, so that by setting $x = u^2$,

$$y(a, t) = \frac{2}{\pi} \int_0^\infty \frac{e^{-tu^2} \sin au}{u}\, du.$$

In view of Theorem A.12 we can differentiate under the integral sign so that

$$\frac{\partial y}{\partial a} = \frac{2}{\pi} \int_0^\infty e^{-tu^2} \cos(au)\, du = \frac{2}{\pi}\, Y(a, t). \tag{A.13}$$

Now,

$$Y(a, t) = \int_0^\infty e^{-tu^2} \cos(au)\, du = \left. \frac{e^{-tu^2} \sin au}{a} \right|_0^\infty$$

$$+ \frac{2t}{a} \int_0^\infty e^{-tu^2} u \sin(au)\, du$$

$$= -\frac{2t}{a} \frac{\partial Y}{\partial a},$$

or

$$\frac{\partial Y}{\partial a} = -\frac{a}{2t}\, Y,$$

where $Y(0, t) = \sqrt{\pi}/2\sqrt{t}$ by (2.49). Solving gives

$$Y(a, t) = \frac{\sqrt{\pi}}{2\sqrt{t}}\, e^{-\frac{a^2}{4t}}.$$

Therefore, by (A.13),

$$\frac{\partial y}{\partial a} = \frac{1}{\sqrt{\pi t}}\, e^{-\frac{a^2}{4t}},$$

and since $y(0, t) = 0$,

$$y(a, t) = \frac{1}{\sqrt{\pi t}} \int_0^a e^{-\frac{w^2}{4t}}\, dw$$

$$= \frac{2}{\sqrt{\pi}} \int_0^{a/2\sqrt{t}} e^{-u^2}\, du$$

$$= \mathrm{erf}\left(\frac{a}{2\sqrt{t}}\right),$$

since we substituted $u^2 = w^2/4t$. $\qquad\square$

Theorem A.14 (Fourier Inversion Theorem). *Suppose that f and f' are piecewise continuous on $(-\infty, \infty)$, that is, both are continuous in any finite interval except possibly for a finite number of jump* *discontinuities. Suppose further that f is absolutely integrable, namely,*

$$\int_{-\infty}^{\infty} |f(t)| \, dt < \infty.$$

Then at each point t where f is continuous,

$$f(t) = \frac{1}{2\pi} \int_{-\infty}^{\infty} e^{i\lambda t} F(\lambda) \, d\lambda, \tag{A.14}$$

where

$$F(\lambda) = \int_{-\infty}^{\infty} e^{-i\lambda t} f(t) \, dt$$

is the Fourier transform of f. At a jump discontinuity t, the integral in (A.14) gives the value

$$\frac{f(t^+) + f(t^-)}{2}.$$

For a proof of this exceptionally important result, see for example, Jerri [6], Theorem 2.14.

References

[1] Ahlfors, L.V., *Complex Analysis*, McGraw-Hill, 3rd ed., 1979.

[2] Doetsch, G., *Introduction to the Theory and Application of the Laplace Transform*, Springer-Verlag, 1970.

[3] Doetsch, G., *Guide to the Applications of Laplace Transforms*, Van Nostrand Co., 1963.

[4] Grove, A.C., *An Introduction to the Laplace Transform and the Z-Transform*, Prentice-Hall, 1991.

[5] Guest, P.B., *Laplace Transforms and an Introduction to Distributions*, Ellis Horwood, 1991.

[6] Jerri, A., *Integral and Discrete Transforms with Applications to Error Analysis*, Marcel Dekker Inc., 1992.

[7] Kuhfittig, P.K.F., *Introduction to the Laplace Transform*, Plenum Press, 1978.

[8] Mickens, R., *Difference Equations*, Van Nostrand Reinhold Co., 1987.

[9] Oberhettinger, F. and Badii, L., *Tables of Laplace Transforms*, Springer-Verlag, 1973.

[10] Protter, M.H. and Morrey, C.B., *A First Course in Real Analysis*, Springer-Verlag, 1977.

[11] Richards, I. and Youn, H., *Theory of Distributions: A Non-Technical Introduction*, Cambridge University Press, 1990.

[12] Roberts, G.E. and Kaufman, H., *Table of Laplace Transforms*, Saunders, 1966.

[13] Schwartz, L., *Théorie des distributions*, Nouvelle Édition, Hermann, 1966.

[14] Watson, E.J., *Laplace Transforms*, Van Nostrand Reinhold Co., 1981.

[15] Widder, D.V., *The Laplace Transform*, Princeton University Press, 1946.

[16] Zill, D.G., *A First Course in Differential Equations with Applications*, 4th ed., 1989, PWS-Kent.

Tables

Laplace Transform Operations

$F(s)$	$f(t)$
$c_1 F_1(s) + c_2 F_2(s)$	$c_1 f_1(t) + c_2 f_2(t)$
$F(as) \quad (a > 0)$	$\dfrac{1}{a} f\left(\dfrac{t}{a}\right)$
$F(s - a)$	$e^{at} f(t)$
$e^{-as} F(s) \quad (a \geq 0)$	$u_a(t) f(t - a)$
$s F(s) - f(0^+)$	$f'(t)$
$s^2 F(s) - s f(0^+) - f'(0^+)$	$f''(t)$
$s^n F(s) - s^{n-1} f(0^+) - s^{n-2} f'(0^+)$ $- \cdots - f^{(n-1)}(0^+)$	$f^{(n)}(t)$
$\dfrac{F(s)}{s}$	$\displaystyle \int_0^t f(\tau)\, d\tau$
$F'(s)$	$-t f(t)$
$F^{(n)}(s)$	$(-1)^n t^n f(t)$
$\displaystyle \int_s^\infty F(x)\, dx$	$\dfrac{1}{t} f(t)$
$F(s)\, G(s)$	$\displaystyle \int_0^t f(\tau) g(t - \tau)\, d\tau$
$\displaystyle \lim_{s \to \infty} s F(s)$	$\displaystyle \lim_{t \to 0^+} f(t) = f(0^+)$
$\displaystyle \lim_{s \to 0} s F(s)$	$\displaystyle \lim_{t \to \infty} f(t)$

Table of Laplace Transforms

$F(s)$	$f(t)$
1	$\delta(t)$
$\dfrac{1}{s}$	1
$\dfrac{1}{s^2}$	t
$\dfrac{1}{s^n} \quad (n = 1, 2, 3, \ldots)$	$\dfrac{t^{n-1}}{(n-1)!}$
$\dfrac{1}{s^v} \quad (v > 0)$	$\dfrac{t^{v-1}}{\Gamma(v)}$
$\dfrac{(s-1)^n}{s^{n+1}} \quad (n = 0, 1, 2, \ldots)$	$L_n(t) = \dfrac{e^t}{n!} \dfrac{d^n}{dt^n} (t^n e^{-t})$ Laguerre polynomials
$\dfrac{1}{s-a}$	e^{at}
$\dfrac{1}{s(s-a)}$	$\dfrac{1}{a}(e^{at} - 1)$
$\dfrac{1}{(s-a)(s-b)} \quad (a \neq b)$	$\dfrac{e^{at} - e^{bt}}{a - b}$
$\dfrac{s}{(s-a)(s-b)} \quad (a \neq b)$	$\dfrac{a\,e^{at} - b\,e^{bt}}{a - b}$
$\dfrac{s}{(s-a)^2}$	$(1 + at)\,e^{at}$
$\dfrac{a}{s^2 + a^2}$	$\sin at$
$\dfrac{s}{s^2 + a^2}$	$\cos at$
$\dfrac{a}{(s-b)^2 + a^2}$	$e^{bt} \sin at$
$\dfrac{s-b}{(s-b)^2 + a^2}$	$e^{bt} \cos at$

$F(s)$	$f(t)$
$\dfrac{a}{s^2 - a^2}$	$\sinh at$
$\dfrac{s}{s^2 - a^2}$	$\cosh at$
$\dfrac{a}{(s-b)^2 - a^2}$	$e^{bt} \sinh at$
$\dfrac{s-b}{(s-b)^2 - a^2}$	$e^{bt} \cosh at$
$\dfrac{1}{(s^2 + a^2)^2}$	$\dfrac{1}{2a^3}(\sin at - at \cos at)$
$\dfrac{s}{(s^2 + a^2)^2}$	$\dfrac{1}{2a}(t \sin at)$
$\dfrac{s^2}{(s^2 + a^2)^2}$	$\dfrac{1}{2a}(\sin at + at \cos at)$
$\dfrac{s^3}{(s^2 + a^2)^2}$	$\cos at - \tfrac{1}{2} at \sin at$
$\dfrac{s^2 - a^2}{(s^2 + a^2)^2}$	$t \cos at$
$\dfrac{1}{(s^2 - a^2)^2}$	$\dfrac{1}{2a^3}(at \cosh at - \sinh at)$
$\dfrac{s}{(s^2 - a^2)^2}$	$\dfrac{1}{2a}(t \sinh at)$

$F(s)$	$f(t)$
$\dfrac{s^2}{(s^2 - a^2)^2}$	$\dfrac{1}{2a}(\sinh at + at\cosh at)$
$\dfrac{s^3}{(s^2 - a^2)^2}$	$\cosh at + \frac{1}{2}\,at\sinh at$
$\dfrac{s^2 + a^2}{(s^2 - a^2)^2}$	$t\cosh at$
$\dfrac{ab}{(s^2 + a^2)(s^2 + b^2)}\quad (a^2 \neq b^2)$	$\dfrac{a\sin bt - b\sin at}{a^2 - b^2}$
$\dfrac{s}{(s^2 + a^2)(s^2 + b^2)}\quad (a^2 \neq b^2)$	$\dfrac{\cos bt - \cos at}{a^2 - b^2}$
$\dfrac{s^2}{(s^2 + a^2)(s^2 + b^2)}\quad (a^2 \neq b^2)$	$\dfrac{a\sin at - b\sin bt}{a^2 - b^2}$
$\dfrac{s^3}{(s^2 + a^2)(s^2 + b^2)}\quad (a^2 \neq b^2)$	$\dfrac{a^2\cos at - b^2\cos bt}{a^2 - b^2}$
$\dfrac{ab}{(s^2 - a^2)(s^2 - b^2)}\quad (a^2 \neq b^2)$	$\dfrac{b\sinh at - a\sinh bt}{a^2 - b^2}$
$\dfrac{s}{(s^2 - a^2)(s^2 - b^2)}\quad (a^2 \neq b^2)$	$\dfrac{\cosh at - \cosh bt}{a^2 - b^2}$
$\dfrac{s^2}{(s^2 - a^2)(s^2 - b^2)}\quad (a^2 \neq b^2)$	$\dfrac{a\sinh at - b\sinh bt}{a^2 - b^2}$
$\dfrac{s^3}{(s^2 - a^2)(s^2 - b^2)}\quad (a^2 \neq b^2)$	$\dfrac{a^2\cosh at - b^2\cosh bt}{a^2 - b^2}$

$F(s)$	$f(t)$
$\dfrac{a^2}{s^2(s^2+a^2)}$	$t - \dfrac{1}{a}\sin at$
$\dfrac{a^2}{s^2(s^2-a^2)}$	$\dfrac{1}{a}\sinh at - t$
$\dfrac{1}{\sqrt{s}}$	$\dfrac{1}{\sqrt{\pi t}}$
$\dfrac{1}{\sqrt{s+a}}$	$\dfrac{e^{-at}}{\sqrt{\pi t}}$
$\dfrac{1}{s\sqrt{s+a}}$	$\dfrac{1}{\sqrt{a}}\,\mathrm{erf}(\sqrt{at})$
$\dfrac{1}{\sqrt{s+a}+\sqrt{s+b}}$	$\dfrac{e^{-bt}-e^{-at}}{2(b-a)\sqrt{\pi t^3}}$
$\dfrac{1}{s\sqrt{s}}$	$2\sqrt{\dfrac{t}{\pi}}$
$\dfrac{1}{(s-a)\sqrt{s}}$	$\dfrac{1}{\sqrt{a}}\,e^{at}\,\mathrm{erf}\,\sqrt{at}$
$\dfrac{1}{\sqrt{s-a}+b}$	$e^{at}\left(\dfrac{1}{\sqrt{\pi t}}-b\,e^{b^2 t}\,\mathrm{erfc}(b\sqrt{t})\right)$
$\dfrac{1}{\sqrt{s^2+a^2}}$	$J_0(at)$
$\dfrac{1}{\sqrt{s^2-a^2}}$	$I_0(at)$
$\dfrac{(\sqrt{s^2+a^2}-s)^\nu}{\sqrt{s^2+a^2}}\quad(\nu>-1)$	$a^\nu J_\nu(at)$
$\dfrac{(s-\sqrt{s^2-a^2})^\nu}{\sqrt{s^2-a^2}}\quad(\nu>-1)$	$a^\nu I_\nu(at)$
$\dfrac{1}{(s^2+a^2)^\nu}\quad(\nu>0)$	$\dfrac{\sqrt{\pi}}{\Gamma(\nu)}\left(\dfrac{t}{2a}\right)^{\nu-\frac{1}{2}}J_{\nu-\frac{1}{2}}(at)$
$\dfrac{1}{(s^2-a^2)^\nu}\quad(\nu>0)$	$\dfrac{\sqrt{\pi}}{\Gamma(\nu)}\left(\dfrac{t}{2a}\right)^{\nu-\frac{1}{2}}I_{\nu-\frac{1}{2}}(at)$

$F(s)$	$f(t)$
$(\sqrt{s^2 + a^2} - s)^\nu \quad (\nu > 0)$	$\dfrac{\nu a^\nu}{t} J_\nu(at)$
$(s - \sqrt{s^2 - a^2})^\nu \quad (\nu > 0)$	$\dfrac{\nu a^\nu}{t} I_\nu(at)$
$\sqrt{s - a} - \sqrt{s - b}$	$\dfrac{1}{2t\sqrt{\pi t}} (e^{bt} - e^{at})$
$\dfrac{e^{-a/s}}{\sqrt{s}}$	$\dfrac{\cos 2\sqrt{at}}{\sqrt{\pi t}}$
$\dfrac{e^{-a/s}}{s\sqrt{s}}$	$\dfrac{\sin 2\sqrt{at}}{\sqrt{\pi a}}$
$\dfrac{e^{-a/s}}{s^{\nu+1}} \quad (\nu > -1)$	$\left(\dfrac{t}{a}\right)^{\nu/2} J_\nu(2\sqrt{at})$
$\dfrac{e^{-a\sqrt{s}}}{\sqrt{s}} \quad (a > 0)$	$\dfrac{e^{-a^2/4t}}{\sqrt{\pi t}}$
$e^{-a\sqrt{s}} \quad (a > 0)$	$\dfrac{a}{2\sqrt{\pi t^3}} e^{-a^2/4t}$
$\dfrac{e^{-a\sqrt{s}}}{s} \quad (a > 0)$	$\text{erfc}\left(\dfrac{a}{2\sqrt{t}}\right)$
$\dfrac{e^{-k\sqrt{s^2+a^2}}}{\sqrt{s^2 + a^2}}$	$\begin{cases} 0 & 0 < t < k \\ J_0(a\sqrt{t^2 - k^2}) & t > k \end{cases}$
$\dfrac{e^{-k\sqrt{s^2-a^2}}}{\sqrt{s^2 - a^2}}$	$\begin{cases} 0 & 0 < t < k \\ I_0(a\sqrt{t^2 - k^2}) & t > k \end{cases}$

$F(s)$	$f(t)$
e^{-as} $(a > 0)$	$\delta_a(t)$
$\dfrac{e^{-as}}{s}$ $(a > 0)$	$u_a(t)$
$e^{s^2/4}\operatorname{erfc}\dfrac{s}{2}$	$\dfrac{2}{\sqrt{\pi}}\,e^{-t^2}$
$\dfrac{1}{s(e^{as}-1)} = \dfrac{e^{-as}}{s(1-e^{-as})}$	$\left[\dfrac{t}{a}\right]$ ([t] : greatest integer $\leq t$)
$\dfrac{1}{s(e^s-a)} = \dfrac{e^{-s}}{s(1-ae^{-s})}$ $(a \neq 1)$	$\dfrac{a^{[t]}-1}{a-1}$
$\dfrac{e^s-1}{s(e^s-a)} = \dfrac{1-e^{-s}}{s(1-ae^{-s})}$	$a^{[t]}$
$\dfrac{1}{s(1-e^{-as})}$	
$\dfrac{1}{s(1+e^{-as})}$	
$\dfrac{1}{s(1+e^{as})}$	
$\dfrac{1-e^{-as}}{s(e^{as}+e^{-as})}$	

$F(s)$	$f(t)$

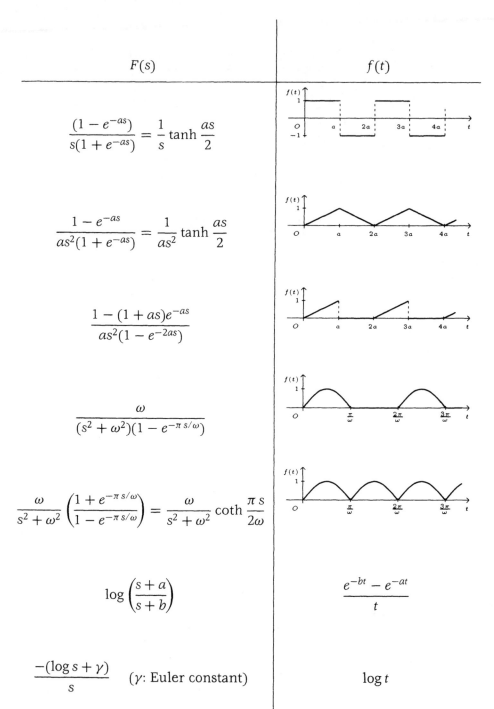

$$\frac{(1 - e^{-as})}{s(1 + e^{-as})} = \frac{1}{s} \tanh \frac{as}{2}$$

$$\frac{1 - e^{-as}}{as^2(1 + e^{-as})} = \frac{1}{as^2} \tanh \frac{as}{2}$$

$$\frac{1 - (1 + as)e^{-as}}{as^2(1 - e^{-2as})}$$

$$\frac{\omega}{(s^2 + \omega^2)(1 - e^{-\pi s/\omega})}$$

$$\frac{\omega}{s^2 + \omega^2}\left(\frac{1 + e^{-\pi s/\omega}}{1 - e^{-\pi s/\omega}}\right) = \frac{\omega}{s^2 + \omega^2} \coth \frac{\pi s}{2\omega}$$

$$\log\left(\frac{s + a}{s + b}\right) \qquad \frac{e^{-bt} - e^{-at}}{t}$$

$$\frac{-(\log s + \gamma)}{s} \quad (\gamma: \text{Euler constant}) \qquad \log t$$

$F(s)$	$f(t)$
$\dfrac{\log s}{s}$	$-(\log t + \gamma)$
$\log\left(\dfrac{s^2 + a^2}{s^2 + b^2}\right)$	$\dfrac{2}{t}(\cos bt - \cos at)$
$\tan^{-1}\left(\dfrac{a}{s}\right)$	$\dfrac{1}{t}\sin at$
$\dfrac{\sinh xs}{s \sinh as}$	$\dfrac{x}{a} + \dfrac{2}{\pi}\displaystyle\sum_{n=1}^{\infty}\dfrac{(-1)^n}{n}\sin\dfrac{n\pi x}{a}\cos\dfrac{n\pi t}{a}$
$\dfrac{\sinh xs}{s \cosh as}$	$\dfrac{4}{\pi}\displaystyle\sum_{n=1}^{\infty}\dfrac{(-1)^{n-1}}{2n-1}\sin\left(\dfrac{2n-1}{2a}\right)\pi x \sin\left(\dfrac{2n-1}{2a}\right)\pi t$
$\dfrac{\cosh xs}{s \sinh as}$	$\dfrac{t}{a} + \dfrac{2}{\pi}\displaystyle\sum_{n=1}^{\infty}\dfrac{(-1)^n}{n}\cos\dfrac{n\pi x}{a}\sin\dfrac{n\pi t}{a}$
$\dfrac{\cosh xs}{s \cosh as}$	$1 + \dfrac{4}{\pi}\displaystyle\sum_{n=1}^{\infty}\dfrac{(-1)^n}{2n-1}\cos\left(\dfrac{2n-1}{2a}\right)\pi x \cos\left(\dfrac{2n-1}{2a}\right)\pi t$
$\dfrac{\sinh xs}{s^2 \sinh as}$	$\dfrac{xt}{a} + \dfrac{2a}{\pi^2}\displaystyle\sum_{n=1}^{\infty}\dfrac{(-1)^n}{n^2}\sin\dfrac{n\pi x}{a}\sin\dfrac{n\pi t}{a}$
$\dfrac{\sinh xs}{s^2 \cosh as}$	$x + \dfrac{8a}{\pi^2}\displaystyle\sum_{n=1}^{\infty}\dfrac{(-1)^n}{(2n-1)^2}\sin\left(\dfrac{2n-1}{2a}\right)\pi x \cos\left(\dfrac{2n-1}{2a}\right)\pi t$
$\dfrac{\cosh xs}{s^2 \sinh as}$	$\dfrac{1}{2a}\left(x^2 + t^2 - \dfrac{a^2}{3}\right) - \dfrac{2a}{\pi^2}\displaystyle\sum_{n=1}^{\infty}\dfrac{(-1)^n}{n^2}\cos\dfrac{n\pi x}{a}\cos\dfrac{n\pi t}{a}$
$\dfrac{\cosh xs}{s^2 \cosh as}$	$t + \dfrac{8a}{\pi^2}\displaystyle\sum_{n=1}^{\infty}\dfrac{(-1)^n}{(2n-1)^2}\cos\left(\dfrac{2n-1}{2a}\right)\pi x \sin\left(\dfrac{2n-1}{2a}\right)\pi t$
$\dfrac{\sinh x\sqrt{s}}{\sinh a\sqrt{s}}$	$\dfrac{2\pi}{a^2}\displaystyle\sum_{n=1}^{\infty}(-1)^n n e^{-n^2\pi^2 t/a^2}\sin\dfrac{n\pi x}{a}$

$F(s)$	$f(t)$
$\dfrac{\cosh x\sqrt{s}}{\cosh a\sqrt{s}}$	$\dfrac{\pi}{a^2}\displaystyle\sum_{n=1}^{\infty}(-1)^{n-1}(2n-1)e^{-(2n-1)^2\pi^2t/4a^2}\cos\left(\dfrac{2n-1}{2a}\right)\pi x$
$\dfrac{\sinh x\sqrt{s}}{\sqrt{s}\cosh a\sqrt{s}}$	$\dfrac{2}{a}\displaystyle\sum_{n=1}^{\infty}(-1)^{n-1}e^{-(2n-1)^2\pi^2t/4a^2}\sin\left(\dfrac{2n-1}{2a}\right)\pi x$
$\dfrac{\cosh x\sqrt{s}}{\sqrt{s}\sinh a\sqrt{s}}$	$\dfrac{1}{a}+\dfrac{2}{a}\displaystyle\sum_{n=1}^{\infty}(-1)^{n}e^{-n^2\pi^2t/a^2}\cos\dfrac{n\pi x}{a}$
$\dfrac{\sinh x\sqrt{s}}{s\sinh a\sqrt{s}}$	$\dfrac{x}{a}+\dfrac{2}{\pi}\displaystyle\sum_{n=1}^{\infty}\dfrac{(-1)^{n}}{n}e^{-n^2\pi^2t/a^2}\sin\dfrac{n\pi x}{a}$
$\dfrac{\cosh x\sqrt{s}}{s\cosh a\sqrt{s}}$	$1+\dfrac{4}{\pi}\displaystyle\sum_{n=1}^{\infty}\dfrac{(-1)^{n}}{2n-1}e^{-(2n-1)^2\pi^2t/4a^2}\cos\left(\dfrac{2n-1}{2a}\right)\pi x$
$\dfrac{\sinh x\sqrt{s}}{s^2\sinh a\sqrt{s}}$	$\dfrac{xt}{a}+\dfrac{2a^2}{\pi^3}\displaystyle\sum_{n=1}^{\infty}\dfrac{(-1)^{n}}{n^3}(1-e^{-n^2\pi^2t/a^2})\sin\dfrac{n\pi x}{a}$
$\dfrac{\cosh x\sqrt{s}}{s^2\cosh a\sqrt{s}}$	$\dfrac{x^2-a^2}{2}+t-\dfrac{16a^2}{\pi^3}\displaystyle\sum_{n=1}^{\infty}\dfrac{(-1)^{n}}{(2n-1)^3}e^{-(2n-1)^2\pi^2t/4a^2}\cos\left(\dfrac{2n-1}{2a}\right)\pi x$

Answers to Exercises

Exercises 1.1.

1. (a) $\dfrac{4}{s^2}$

(b) $\dfrac{1}{s-2}$

(c) $\dfrac{2s}{s^2+9}$

(d) $\dfrac{1}{s} - \dfrac{s}{s^2+\omega^2}$

(e) $-\dfrac{1}{(s-2)^2}$

(f) $\dfrac{1}{s^2-2s+2}$

(g) $\dfrac{e^{-as}}{s}$

(h) $\dfrac{\omega(1+e^{-\pi s/\omega})}{s^2+\omega^2}$

(i) $\dfrac{2}{s} - \dfrac{2e^{-s}}{s} + \dfrac{e^{-(s-1)}}{s-1}$

2. (a) $\dfrac{1}{s}\left(\dfrac{e^{-s}}{s} - \dfrac{1}{s} + 1\right)$

(b) $\dfrac{1}{s^2}(1-e^{-s})^2$

Exercises 1.3.

1. $f(t)$ is continuous except at $t = -1$.

2. $g(t)$ is continuous on $\mathbb{R}\backslash\{0\}$, and also at 0 if we define $g(0) = 0$.

3. $h(t)$ is continuous on $\mathbb{R}\backslash\{1\}$, with a jump discontinuity at $t = 1$.

4. $i(t)$ is continuous on $\mathbb{R}$.

5. $j(t)$ is continuous on $\mathbb{R}\backslash\{0\}$.

6. $k(t)$ is continuous on $\mathbb{R}\backslash\{0\}$, with a jump discontinuity at $t = 0$.

7. $l(t)$ is continuous except at the points $t = a, 2a, 3a, \ldots$, where it has a jump discontinuity.

8. $m(t)$ is continuous except at the points $t = a, 2a, 3a, \ldots$, where it has a jump discontinuity.

Exercises 1.4.

1. (i) $c_1 f_1 + c_2 f_2$ is piecewise continuous, of order $\max(\alpha, \beta)$.
   (ii) $f \cdot g$ is piecewise continuous, of order $\alpha + \beta$.

Exercises 1.5.

1. (a) Yes.   No.

Exercises 1.6.

1. $\dfrac{2}{s^2} + \dfrac{3}{(s-2)} + \dfrac{12}{s^2+9}$

3. (a) $\dfrac{s^2 - 2\omega^2}{s(s^2 - 4\omega^2)}$ **(b)** $\dfrac{2\omega^2}{s(s^2 - 4\omega^2)}$

4. $\dfrac{3s - 4}{s^2 - 4}$

5. $\displaystyle\sum_{n=0}^{\infty} \dfrac{(-1)^n \omega^{2n}}{s^{2n+1}} = \dfrac{s}{s^2 + \omega^2}$, $\displaystyle\sum_{n=0}^{\infty} \dfrac{(-1)^n \omega^{2n+1}}{s^{2n+2}} = \dfrac{\omega}{s^2 + \omega^2}$

6. $\dfrac{2\omega^2}{s(s^2 + 4\omega^2)}$, $\dfrac{s^2 + 2\omega^2}{s(s^2 + 4\omega^2)}$

7. $\log\left(1 + \dfrac{1}{s}\right)$

8. $\displaystyle\sum_{n=1}^{\infty} \dfrac{(-1)^{n+1}}{2n} \left(\dfrac{\omega}{s}\right)^{2n} = \dfrac{1}{2}\log\left(1 + \dfrac{\omega^2}{s^2}\right)$

9. No.

Exercises 1.7.

2. (a) $N(t) = \begin{cases} 1 & t = 0 \\ 0 & t \neq 0 \end{cases}$

(There are many other examples.)

(d) $f(t) \equiv 0$ is the only continuous null function.

3. (b) $f(t) = \sum_{n=0}^{\infty} u_{na}(t)$.

Exercises 1.8.

1. (a) $\dfrac{3}{(s-2)^2 + 9}$

(b) $\dfrac{2}{(s+\omega)^3}$

(c) $2t^2 e^{4t}$

(d) $\dfrac{\sqrt{2}}{(s-7)^2 - 2}$

(e) $\frac{1}{2} e^{-t} \sin 2t$

(f) $e^{-3t} \left(\cosh(2\sqrt{2}\,t) - \dfrac{3}{2\sqrt{2}} \sinh(2\sqrt{2}\,t) \right)$

(g) $\dfrac{(\cos\theta)(s+a) - (\sin\theta)\omega}{(s+a)^2 + \omega^2}$

(h) $e^{-t}(1 - t)$

2. (a) $\dfrac{e^{-2(s-a)}}{s-a}$

(b) $\dfrac{s}{(s^2+1)} e^{-\pi s/2}$

(c) $\dfrac{s}{s^2+1} e^{-\pi s}$

3. (a) $\frac{1}{2} u_2(t)(t-2)^2$

(b) $E - u_a(t) \cos(t-a)$

(c) $\dfrac{1}{\sqrt{2}} u_\pi(t) \sinh\left(\sqrt{2}(t-\pi)\right)$

Exercises 1.9.

1. (a) $\dfrac{s^2 + \omega^2}{(s^2 - \omega^2)^2}$

(b) $\dfrac{2\omega s}{(s^2 - \omega^2)^2}$

(c) $\dfrac{2s(s^2 - 3\omega^2)}{(s^2 + \omega^2)^3}$

(d) $\dfrac{2\omega(2s^2 - \omega^2)}{(s^2 + \omega^2)^3}$

3. (a) $\dfrac{2}{t}(\cos bt - \cos at)$ (b) $\dfrac{\sin t}{t}$

4. $\dfrac{ae^{-a^2/4t}}{2t\sqrt{\pi t}}$

Exercises 1.10.

1. (a) $\dfrac{e^{at} - e^{bt}}{a - b}$ (b) $\frac{1}{3}e^{-t} + \frac{1}{6}e^{t/2}$

(c) $-1 + e^t + t^2 e^t$ (d) $\dfrac{1}{b^2 - a^2}(\cos at - \cos bt)$

(e) $\dfrac{1}{a^2 + b^2}(\cosh bt - \cos at)$ (f) $\frac{5}{2} + 2t + \frac{t^2}{2} - 3e^t + \frac{1}{2}e^{2t}$

(g) $\frac{3}{2}e^{-t} + \frac{5}{4}te^{-t} - \frac{3}{2}\cos t + \frac{1}{4}\sin t - \frac{1}{4}t\sin t$

(h) $\frac{1}{4} - \frac{5}{4}e^{3t} + \frac{23}{20}e^{4t} - \frac{3}{20}e^{-t}$

2. The answer for both parts (a) and (b) is

$$\dfrac{a \sinh at}{(a^2 - b^2)(a^2 - c^2)} + \dfrac{b \sinh bt}{(b^2 - a^2)(b^2 - c^2)} + \dfrac{c \sinh ct}{(c^2 - a^2)(c^2 - b^2)}.$$

Exercises 2.1.

2. (a) $\dfrac{\sqrt{\pi}}{2}$ (b) 2 (c) $-2\sqrt{\pi}$ (d) $\dfrac{4\sqrt{\pi}}{3}$

3. (a) $\dfrac{\sqrt{\pi}}{\sqrt{s - 3}}$ (b) $\dfrac{u_2(t)}{\sqrt{\pi(t - 2)}}$

(c) $\dfrac{2}{\sqrt{\pi}}t^{1/2}e^{at}$ (d) e^{-t}

(e) $\displaystyle\sum_{n=1}^{\infty} \dfrac{(-1)^{n+1}t^{2n-1}}{n(2n - 1)!} = 2\left(\dfrac{1 - \cos t}{t}\right)$

(f) $\dfrac{\sqrt{\pi}}{2s^{3/2}}$

Exercises 2.2.

1. (a) $\dfrac{1}{s(1 + e^{-as})}$

(b) $\dfrac{1}{s}\left(\dfrac{1-e^{-as}}{1+e^{-as}}\right) = \dfrac{1}{s}\tanh\dfrac{as}{2}$

(c) $\dfrac{1-e^{-as}-as\,e^{-as}}{a\,s^2(1-e^{-2as})}$

(d) $\dfrac{1}{as^2}\left(\dfrac{1-e^{-as}}{1+e^{-as}}\right) = \dfrac{1}{as^2}\tanh\dfrac{as}{2}$

2. $\dfrac{1}{s(1+e^{-as})}$

3. $f(t) = u(t) + 2\sum_{n=1}^{\infty}(-1)^n u_{na}(t)$

$F(s) = \dfrac{1}{s}\tanh\dfrac{as}{2}$

4. $F(s) = \dfrac{1}{s}\sum_{n=0}^{\infty}(-1)^n(e^{-as(2n+1)} - e^{-2as(n+1)})$

$f(t) = \sum_{n=0}^{\infty}(-1)^n\left(u_{(2n+1)a}(t) - u_{2a(n+1)}(t)\right)$

Graph of $f(t)$:

Exercises 2.3.

5. (a) $\dfrac{6\omega^3}{(s^2+9\omega^2)(s^2+\omega^2)}$

(b) $\dfrac{s(s^2+7\omega^2)}{(s^2+9\omega^2)(s^2+\omega^2)}$

6. Use induction.

7. $f(t) = e^{[t+1]^2}$, where $[t]$ = greatest integer $\le t$.

Exercises 2.4.

1. (a) $y = -\frac{1}{2}(e^t + \cos t - \sin t)$

(b) $y = e^t\left(\frac{1}{2}t^2 - \frac{1}{2}t + \frac{1}{4}\right) + \frac{7}{4}e^{-t}$

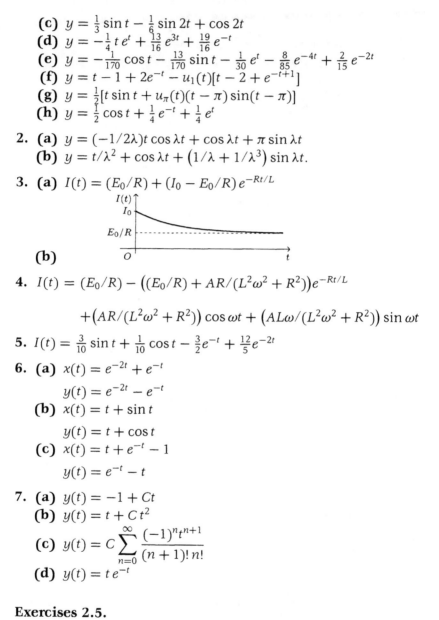

(c) $y = \frac{1}{3}\sin t - \frac{1}{6}\sin 2t + \cos 2t$

(d) $y = -\frac{1}{4}te^t + \frac{13}{16}e^{3t} + \frac{19}{16}e^{-t}$

(e) $y = -\frac{1}{170}\cos t - \frac{13}{170}\sin t - \frac{1}{30}e^t - \frac{8}{85}e^{-4t} + \frac{2}{15}e^{-2t}$

(f) $y = t - 1 + 2e^{-t} - u_1(t)[t - 2 + e^{-t+1}]$

(g) $y = \frac{1}{2}[t\sin t + u_\pi(t)(t - \pi)\sin(t - \pi)]$

(h) $y = \frac{1}{2}\cos t + \frac{1}{4}e^{-t} + \frac{1}{4}e^t$

2. (a) $y = (-1/2\lambda)t\cos\lambda t + \cos\lambda t + \pi\sin\lambda t$

(b) $y = t/\lambda^2 + \cos\lambda t + (1/\lambda + 1/\lambda^3)\sin\lambda t.$

3. (a) $I(t) = (E_0/R) + (I_0 - E_0/R)e^{-Rt/L}$

(b)

4. $I(t) = (E_0/R) - ((E_0/R) + AR/(L^2\omega^2 + R^2))e^{-Rt/L}$

$$+ (AR/(L^2\omega^2 + R^2))\cos\omega t + (AL\omega/(L^2\omega^2 + R^2))\sin\omega t$$

5. $I(t) = \frac{3}{10}\sin t + \frac{1}{10}\cos t - \frac{3}{2}e^{-t} + \frac{12}{5}e^{-2t}$

6. (a) $x(t) = e^{-2t} + e^{-t}$

$y(t) = e^{-2t} - e^{-t}$

(b) $x(t) = t + \sin t$

$y(t) = t + \cos t$

(c) $x(t) = t + e^{-t} - 1$

$y(t) = e^{-t} - t$

7. (a) $y(t) = -1 + Ct$

(b) $y(t) = t + Ct^2$

(c) $y(t) = C\sum_{n=0}^{\infty}\frac{(-1)^n t^{n+1}}{(n+1)!\,n!}$

(d) $y(t) = te^{-t}$

Exercises 2.5.

1. $y(t) = \frac{e^{2t}}{\sqrt{2}}\sinh\sqrt{2}\,t$

2. $x(t) = (1/\sqrt{km})\sin\left(\sqrt{k/m}\,t\right)$

3. $I(t) = (1/L)e^{-Rt/L}$

4. $x(t) = t\,e^{-t}$

Exercises 2.6.

1. (a) 0

(b) $f(0^+) = \begin{cases} 1 & \text{if } n = 0 \\ 0 & \text{if } n > 0 \end{cases}$

(c) $a - b$

2. (a) 0

(b) 1

Exercises 2.7.

1. (a) $\frac{1}{3}(e^t - e^{-2t})$ **(b)** $1 - \cos t$ **(c)** $t - \sin t$

(d) $\frac{1}{32}(2t - 1) + \frac{e^{-4t}}{32}(2t + 1)$

(e) $\frac{1}{8}(t \sin t - t^2 \cos t)$

6. (a) $(1/\sqrt{a})\,\mathrm{erf}(\sqrt{at})$

(b) $\dfrac{\sqrt{a}}{(s - a)\sqrt{s}}$

(c) $\dfrac{\sqrt{a}(3s + 2a)}{2s^2(s + a)^{3/2}}$

7. (a) $u_1(t)J_0(t - 1)$

(b) $\displaystyle\sum_{n=0}^{\infty} \frac{(-1)^n a^{2n} t^{2n+1}}{2^{2n}(2n + 1)(n!)^2}$

8. $\pi/2$

10. (a) $1 + \frac{2}{\sqrt{3}} \sin\left(\frac{\sqrt{3}}{2}t\right) e^{t/2}$

(b) $-\frac{1}{5}\cos t + \frac{3}{5}\sin t + \frac{1}{5}e^{2t}$

(c) 0

(d) e^{-at}

11. (a) $\frac{1}{2}(\sin t + t \cos t)$

(b) Same as for 7(b) with $a = 1$

12. $\frac{1}{4}\int_0^t (e^{3\tau} - e^{-\tau}) f(t - \tau)\,d\tau$

Exercises 2.8.

1. (a) $\frac{2}{3}e^t + \frac{4}{3}e^{-2t} - 2e^{-t}$

(b) $\frac{1}{4}e^t - \frac{1}{4}e^{-t} - \frac{1}{2}\sin t$

(c) $-\frac{1}{3}e^t + \frac{1}{30}e^{-2t} + \frac{1}{45}e^{3t} + \frac{1}{3}t + \frac{5}{18}$

4. $\frac{1}{2}e^{-t/2}\left(\cos\dfrac{\sqrt{7}}{2}t - \dfrac{1}{\sqrt{7}}\sin\dfrac{\sqrt{7}}{2}t\right) + \frac{1}{2}(\sin t - \cos t)$

Exercises 2.9.

4. (a) $a_n = 3^n - 4^n$ **(b)** $a_n = \frac{1}{2}(2^n - 4^n)$

    (c) $a_n = \frac{1}{2}[1 - (-1)^n]$ **(d)** $a_n = n$

6. (a) $y(t) = \sum_{n=0}^{[t]}(-1)^n e^{t-n}$ **(b)** $y(t) = \sum_{n=0}^{[t]}(t - n)^{n+2}/(n + 2)!$

7. $a_n = 4 + 2n - 7 \cdot 2^n + 3^{n+1}$

Exercises 3.1.

1. (a) $8 + i$ **(b)** $24 + 18i$ **(c)** $\frac{7}{5} - \frac{4}{5}i$

2. (a) $|(1 + i)^3| = 2\sqrt{2}, \quad \arg((1 + i)^3) = 3\pi/4$

       $\mathcal{R}e((1 + i)^3) = -2, \quad \mathcal{I}m((1 + i)^3) = 2$

   (b) $\left|\dfrac{1 - i}{1 + i}\right| = 1, \quad \arg\left(\dfrac{1 - i}{1 + i}\right) = \dfrac{3\pi}{2}$

       $\mathcal{R}e\left(\dfrac{1 - i}{1 + i}\right) = 0, \quad \mathcal{I}m\left(\dfrac{1 - i}{1 + i}\right) = -1$

   (c) $\left|\dfrac{1}{(1 - i)^2}\right| = \dfrac{1}{2}, \quad \arg\left(\dfrac{1}{(1 - i)^2}\right) = \dfrac{\pi}{2}$

       $\mathcal{R}e\left(\dfrac{1}{(1 - i)^2}\right) = 0, \quad \mathcal{I}m\left(\dfrac{1}{(1 - i)^2}\right) = \dfrac{1}{2}$

   (d) $\left|\dfrac{4 + 3i}{2 - i}\right| = \sqrt{5}, \quad \arg\left(\dfrac{4 + 3i}{2 - i}\right) = \tan^{-1}(2)$

       $\mathcal{R}e\left(\dfrac{4 + 3i}{2 - i}\right) = 1, \quad \mathcal{I}m\left(\dfrac{4 + 3i}{2 - i}\right) = 2$

   (e) $|(1 + i)^{30}| = 2^{15}, \quad \arg((1 + i)^{30}) = \dfrac{3\pi}{2}$

       $\mathcal{R}e((1 + i)^{30}) = 0, \quad \mathcal{I}m((1 + i)^{30}) = 2^{15}.$

3. (a) $(1 + i)^3 = 2\sqrt{2}\,e^{i\frac{3\pi}{4}}$ **(d)** $(4 + 3i)/(2 - i) = \sqrt{5}\,e^{i\tan^{-1}(2)}$

7. (a) $|z - i| < 1$ **(b)** $1 \le |z| \le 2$

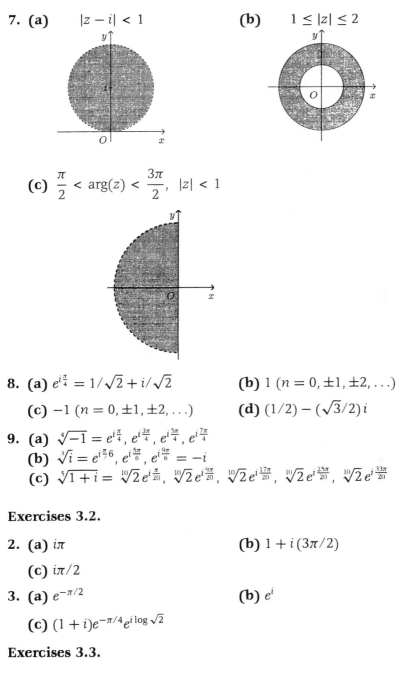

(c) $\dfrac{\pi}{2} < \arg(z) < \dfrac{3\pi}{2}, \ |z| < 1$

8. (a) $e^{i\frac{\pi}{4}} = 1/\sqrt{2} + i/\sqrt{2}$ **(b)** $1 \ (n = 0, \pm 1, \pm 2, \ldots)$

    (c) $-1 \ (n = 0, \pm 1, \pm 2, \ldots)$ **(d)** $(1/2) - (\sqrt{3}/2)i$

9. (a) $\sqrt[4]{-1} = e^{i\frac{\pi}{4}}, \ e^{i\frac{3\pi}{4}}, \ e^{i\frac{5\pi}{4}}, \ e^{i\frac{7\pi}{4}}$

    (b) $\sqrt[3]{i} = e^{i\frac{\pi}{6}}, \ e^{i\frac{5\pi}{6}}, \ e^{i\frac{9\pi}{6}} = -i$

    (c) $\sqrt[5]{1+i} = \sqrt[10]{2}\, e^{i\frac{\pi}{20}}, \ \sqrt[10]{2}\, e^{i\frac{9\pi}{20}}, \ \sqrt[10]{2}\, e^{i\frac{17\pi}{20}}, \ \sqrt[10]{2}\, e^{i\frac{25\pi}{20}}, \ \sqrt[10]{2}\, e^{i\frac{33\pi}{20}}$

Exercises 3.2.

2. (a) $i\pi$ **(b)** $1 + i\,(3\pi/2)$

    (c) $i\pi/2$

3. (a) $e^{-\pi/2}$ **(b)** e^i

    (c) $(1 + i)e^{-\pi/4}e^{i\log\sqrt{2}}$

Exercises 3.3.

1. (a) $2\pi i$ **(b)** $2\pi i$ **(c)** 0

(d) $i\pi$ **(e)** 0 **(f)** 0

(g) $2\pi i(2 - \cos 1)$ **(h)** $\pi i(-24\pi^2 + 6)$

2. (a) $-i\pi$ **(b)** $\left(1 + (\pi/2)\right) + i(1 - \pi)$

5. $|f^{(4)}(0)| \leq 120$

6. Look at $1/f(z)$.

7. Look at $f(z)/e^z$.

Exercises 3.4.

1. (a) $R = 1$ **(b)** $R = \infty$

(c) $R = 1$ **(d)** $R = \infty$

2. (a) $e^{z^2} = \sum_{n=0}^{\infty} z^{2n}/n!$, $\quad R = \infty$

(b) $\sinh z = \sum_{n=0}^{\infty} z^{2n+1}/(2n + 1)!$, $\quad R = \infty$

(c) $1/(1 - z) = \sum_{n=0}^{\infty} z^n$, $\quad R = 1$

(d) $\log(1 + z) = \sum_{n=0}^{\infty} (-1)^n z^{n+1}/(n + 1)$, $\quad R = 1$

4. (a) $z = 0$ (simple pole)

$\quad z = \pm i$ (poles of order 2)

(b) $z = 0$ (pole of order 3)

(c) $z = 0$ (essential singularity)

(d) $z = 1$ (removable singularity)

5. (a) $1 + \frac{z^2}{6} + \frac{7z^4}{360}$ **(b)** $\frac{1}{z^2} - \frac{1}{6} + \frac{7z^2}{360}$

(c) $1 - \frac{z}{3} + \frac{2z^2}{15}$

6. (a) $-\frac{1}{3z} + \frac{1}{4}\sum_{n=0}^{\infty}(-1)^n z^n - \frac{1}{4}\sum_{n=0}^{\infty} z^n/3^{n+2}$

(b) $-\frac{1}{3z} + \frac{1}{4}\sum_{n=0}^{\infty} (-1)^n/z^{n+1} - \frac{1}{4}\sum_{n=0}^{\infty} z^n/3^{n+2}$

(c) $-\frac{1}{3z} + \frac{1}{4}\sum_{n=0}^{\infty} (-1)^n/z^{n+1} + \frac{1}{12}\sum_{n=0}^{\infty} 3^n/z^{n+1}$

7. (a) $\mathrm{Res}(\pm ia) = 1/2$

(b) $\mathrm{Res}(0) = 1/2$

$$\mathrm{Res}\left(\frac{(2n - 1)}{a}\pi i\right) = \frac{i}{(2n - 1)\pi} \quad n = 0, \pm 1, \pm 2, \ldots$$

(c) $\mathrm{Res}(0) = 0$

8. (a) $-2\pi i$ **(b)** 0 **(c)** $6\pi i$

(d) $-4\pi i$ **(e)** 0

9. $-\dfrac{2\pi}{7}$

Exercises 4.

1. (a) $\left(1/(a-b)\right)(a\,e^{at} - b\,e^{bt})$ **(b)** $e^{at}\left(\frac{1}{2}at^2 + t\right)$

(c) $\frac{1}{2a}t\sin at$ **(d)** $t\cosh at$

(e) $\frac{1}{8a}(3t\sin at + at^2\cos at)$

6. $1/\sqrt{s+1}(\sqrt{s+1}+1)$

Exercises 5.

1. $y(x,t) = x(t - 1 + 2e^{-t})$

2. (a) $u(x,t) = (x/2\sqrt{\pi t^3})\,e^{-x^2/4t}$

(b) $u(x,t) = u_0 + (u_1 - u_0)\,\mathrm{erfc}\left(x/2\sqrt{t}\right)$

(c) $u(x,t) = x + (2/\pi)\sum_{n=1}^{\infty}\left((-1)^n/n\right)e^{-n^2\pi^2 t}\sin n\pi x$

(d) $u(x,t) = (2a\ell/\pi)\sum_{n=1}^{\infty}\left((-1)^{n+1}/n\right)e^{-n^2\pi^2 t/\ell^2}\sin(n\pi x/\ell)$

3. (a) $y(x,t) = \sin\pi x\cos\pi t$

(b) $y(x,t) = 1 + (4/\pi)\sum_{n=1}^{\infty}\dfrac{(-1)^n}{2n-1}\cos\left(\frac{2n-1}{2}\right)\pi x\cos\left(\frac{2n-1}{2}\right)\pi t$

(c) $y(x,t) = (2/\pi^2)\sum_{n=1}^{\infty}\left((-1)^{n+1}/n^2\right)\sin n\pi x\sin n\pi t$

(d) $y(x,t) = 2\sum_{n=1}^{\infty}\left(\int_0^1 f(u)\sin n\pi u\,du\right)\sin n\pi x\cos n\pi t$

4. $y(x,t) = \left((\sin\pi x)/\pi^2\right)(\cos\pi t - 1)$

5. $u(x,t) = (1/a\sqrt{\pi t})e^{-x^2/4a^2 t}$

Index

Analytic functions, 123
Argument, 117
Asymptotic values, 88

Bessel function, 72, 97, 213, 214
Beta function, 96
Boundary-value problems, 64
Branch
  cut, 122
  point, 123, 167
Bromwich
  line, 152.
  contour, 152

Cauchy
  inequality, 134, 145
  integral formula, 133
  residue theorem, 143
  –Riemann equations, 123
  theorem, 131
Circle of convergence, 137
Closed (contour), 128

Complex
  inversion formula, 151
  numbers, 115
  plane, 117
Complementary error function,
    172
Conjugate, 116
Continuity, 8
  piecewise, 10
Contour, 128
Convergence, 2, 6
  absolute, 6
  uniform, 7, 20
Cycloid, 101

De Moivre's theorem, 117
Derivative theorem, 54
Difference equations, 108
Differential equations, 59
Differentiation
  of Laplace transform, 31
  under the integral sign, 203

231

Diffusivity, 180
Divergence, 2
Dirac operator, 74, 210, 215

Electrical circuits, 68, 83
Elliptic equations, 175
Equation of motion, 85
Error function, 95
Euler
    constant, 44, 47
    formula, 3, 117
Excitation, 61
Exponential order, 12

Fibonacci equation, 114
First translation theorem, 27
Forcing function, 61
Fourier
    inversion formula, 205
    series, 163
    transform, 151
Full-wave-rectified sine, 51, 216
Fundamental theorem of
        algebra, 200
Functions (complex-valued),
        120

Gamma function, 41
General solutions, 63
Greatest integer $\leq t$, 109, 113,
        215

Half-wave-rectified sine, 50,
        216
Harmonic
    function, 126
    conjugate, 126
Heat equation, 175, 180
Heaviside
    expansion theorem, 107

function, 25, 79, 215
Hyperbolic
    equations, 175
    functions, 121

Impulsive response, 104
Imaginary
    axis, 117
    number, 116
    part, 116
Independence of path, 132
Indicial response, 105
Infinite series, 17, 44
Initial
    point, 128
    -value theorem, 88
Input, 61
Integral equations, 98
Integrals, 66
Integration, 33, 128
Integro-differential equations,
        67

Jump discontinuity, 8

Kirchoff's voltage law, 68

Laplace
    operator, 126
    transform (definition), 1, 78
    transform method, 60, 176
    transform tables, 210
    -Stieltjes transform, 78
Laurent series, 139
Lerch's theorem, 24
Linearity, 16
Liouville's theorem, 134
Logarithm, 122, 216, 217

Mechanical system, 84

Meromorphic function, 141
Modified bessel function, 102,
   213, 214
Modulus, 116
Multiple-valued function, 120

Null function, 26

One-dimensional
  heat equation, 180
  wave equation, 186
Order (of a pole), 141
Ordinary differential equations,
  59
  with polynomial coefficients,
   70
Output, 61

Parabolic equations, 175
Partial
  differential equations, 175
  fractions, 35
Partition, 75, 193
Periodic functions, 47
Positive direction, 128
Polar Form, 117
Pole, 141
Power series, 136
Principal logarithm, 122

Radius of convergence, 136
Real part, 116
Residue, 38, 142
Response, 61
Riemann
  integrable, 193
  integral, 194

–Stieltjes integral, 75, 201
Roots
  of unity, 118
  of a complex number, 118

Second translation theorem, 29
Simple
  contour, 128
  pole, 38, 141
Simply connected, 130
Sine integral, 67
Single-valued functions, 120
Singularities
  essential, 141
  pole, 141
  removable, 141
Smooth (contour), 128
Square–wave, 49, 215
Steady-state solutions, 103
Systems of differential
  equations, 65
Superposition principle, 106

Tautochrone, 100
Taylor
  coefficients, 138
  series, 138
Terminal
  point, 128
  -value theorem, 89
Translation theorems, 27

Uniqueness of inverse, 23
Unit step function, 24, 79, 215

Wave equation, 176, 186

Isaac: The Pleasures of Probability.
Readings in Mathematics.
James: Topological and Uniform
Spaces.
Jänich: Linear Algebra.
Jänich: Topology.
Kemeny/Snell: Finite Markov Chains.
Kinsey: Topology of Surfaces.
Klambauer: Aspects of Calculus.
Lang: A First Course in Calculus. Fifth
edition.
Lang: Calculus of Several Variables.
Third edition.
Lang: Introduction to Linear Algebra.
Second edition.
Lang: Linear Algebra. Third edition.
Lang: Undergraduate Algebra. Second
edition.
Lang: Undergraduate Analysis.
Lax/Burstein/Lax: Calculus with
Applications and Computing.
Volume 1.
LeCuyer: College Mathematics with
APL.
Lidl/Pilz: Applied Abstract Algebra.
Second edition.
Logan: Applied Partial Differential
Equations.
Macki-Strauss: Introduction to Optimal
Control Theory.
Malitz: Introduction to Mathematical
Logic.
Marsden/Weinstein: Calculus I, II, III.
Second edition.
Martin: The Foundations of Geometry
and the Non-Euclidean Plane.
Martin: Geometric Constructions.
Martin: Transformation Geometry: An
Introduction to Symmetry.
Millman/Parker: Geometry: A Metric
Approach with Models. Second
edition.
Moschovakis: Notes on Set Theory.
Owen: A First Course in the
Mathematical Foundations of
Thermodynamics.
Palka: An Introduction to Complex
Function Theory.

Pedrick: A First Course in Analysis.
Peressini/Sullivan/Uhl: The Mathematics
of Nonlinear Programming.
Prenowitz/Jantosciak: Join Geometries.
Priestley: Calculus: A Liberal Art.
Second edition.
Protter/Morrey: A First Course in Real
Analysis. Second edition.
Protter/Morrey: Intermediate Calculus.
Second edition.
Roman: An Introduction to Coding and
Information Theory.
Ross: Elementary Analysis: The Theory
of Calculus.
Samuel: Projective Geometry.
Readings in Mathematics.
Scharlau/Opolka: From Fermat to
Minkowski.
Schiff: The Laplace Transform: Theory
and Applications.
Sethuraman: Rings, Fields, and Vector
Spaces: An Approach to Geometric
Constructability.
Sigler: Algebra.
Silverman/Tate: Rational Points on
Elliptic Curves.
Simmonds: A Brief on Tensor Analysis.
Second edition.
Singer: Geometry: Plane and Fancy.
Singer/Thorpe: Lecture Notes on
Elementary Topology and
Geometry.
Smith: Linear Algebra. Third edition.
Smith: Primer of Modern Analysis.
Second edition.
Stanton/White: Constructive
Combinatorics.
Stillwell: Elements of Algebra:
Geometry, Numbers, Equations.
Stillwell: Mathematics and Its History.
Stillwell: Numbers and Geometry.
Readings in Mathematics.
Strayer: Linear Programming and Its
Applications.
Thorpe: Elementary Topics in
Differential Geometry.
Toth: Glimpses of Algebra and
Geometry.
Readings in Mathematics.

Troutman: Variational Calculus and Optimal Control. Second edition.

Valenza: Linear Algebra: An Introduction to Abstract Mathematics.

Whyburn/Duda: Dynamic Topology.

Wilson: Much Ado About Calculus.

9781475772623